PRINCIPLES OF ELECTRIC VEHICLE TECH

Preface

Welcome to the Principles of Electric Vehicle Technology

As the demand for fuel rises and environmental pollution worsens, finding alternatives to traditional vehicle propulsion methods becomes crucial. Electric vehicles (EVs) are not just a trend, they form a key component of current and future transportation. EVs offer many benefits over conventional vehicles, such as lower emissions, lower fuel costs, and lower maintenance costs. However, EVs also pose new challenges and opportunities for anyone involved or associated with the automotive industry.

It is essential to embed electric vehicle technologies in automotive qualifications and to integrate them at all levels and subjects. Whether you are a vehicle technician, automotive trainer, student, or a member of the emergency services, staying informed about current and emerging propulsion sources is vital for safe interaction with these vehicles. You will also discover the benefits and challenges of EV technology, and how it can transform the way we travel and live.

Please note working on high-voltage vehicle systems without appropriate training can lead to injury or death. This book is not a substitute for training. Always take necessary precautions when working with or around electric vehicles.

This book seeks to bridge the gap between traditional propulsion methods and electric vehicles. It explores why alternative propulsion and electric vehicles are needed and the problems caused by continued fossil fuel usage. It provides a background on electrical fundamentals related to the operation and maintenance of electric vehicles. It outlines some organisational requirements, such as resources, tooling, and equipment needed for maintaining and repairing electric vehicles in workshops. It also introduces the components of electric vehicle drive and operating systems, as well as the maintenance, repair, and diagnosis of these systems.

Key terms, points of interest, safety warnings, and diagnostic tips are provided throughout the text to support the information presented.

Chapters:

How to use this book........................**Page 3**

Chapter 1 Electrical Fundamentals........**Page 5**

Chapter 2 Hybrid and Electric Vehicles..**Page 38**

Chapter 3 Electric Vehicle Hazards and Maintenance......................................**Page 72**

Chapter 4 High-Voltage Component Replacement..................................**Page 107**

Chapter 5 High-Voltage Diagnosis and Testing..**Page 155**

This book offers:

Comprehensive information for students and instructors involved in electric vehicle automotive courses.

Content that supports the knowledge requirements for Levels 1 to 4 EV qualifications.

Numerous illustrations and images that enhance understanding and learning.

Text © Graham Stoakes 2024

Original illustrations and photographs © Graham Stoakes 2024

The rights of Graham Stoakes to be identified as author of this work have been asserted by them in accordance with the Copyright, Designs and Patents Act 1988.

PRINCIPLES OF ELECTRIC VEHICLE TECHNOLOGY

Copyright notice ©

All rights reserved. No part of this publication may be reproduced in any form or by any means (including photocopying or storing it in any medium by electronic means and whether or not transiently or incidentally to some other use of this publication) without the written permission of the copyright owner, except in accordance with the provisions of the Copyright, Designs and Patents Act 1988 or under the terms of a licence issued by the Copyright Licensing Agency, Saffron House, 6 - 10 Kirby Street, London EC1N 8TS (www.cla.co.uk). Applications for the copyright owners' written permission should be addressed to the author.

Acknowledgements

Graham Stoakes would like to thank Anita and Holly Stoakes for their support during this project.

Thank you to alerrandre for the cover design.

The author and publisher would also like to thank the following individuals and organisations for permission to reproduce photographs:

ProMoto Europe

RWC Training

Cover image: Shutterstock.com - UKRID

Author

Graham Stoakes AAE MIMI QTLS is a trainer/lecturer and author of college textbooks in automotive engineering for light vehicles and motorcycles.

With his background as a qualified Master Technician, senior automotive manager, and specialist diagnostic trainer, he brings over 40 years of technical industry experience to this title.

www.grahamstoakes.com

Cover design - fiverr.com/alerrandre

Published by Graham Stoakes

First published 2024

First edition

ISBN 978-0-9929492-7-3

Introduction

How to use this book

This book provides current information on vehicle technologies, but you should keep in mind that technology is always evolving. It's designed to help those who work on, or plan to work on, electric/hybrid vehicles. It provides the knowledge and skills needed to set up and manage resources for electric vehicle repairs. However, it should not be considered a substitute for quality professional training.

To enhance comprehension, the chapters progressively delve deeper into the subject matter, with an increase in topic complexity and content level. However, all chapters are designed to be used in conjunction with each other to provide a comprehensive coverage of knowledge and understanding.

Electric vehicles have high-voltage systems that can be dangerous, especially during maintenance and repair when safety systems may be bypassed or disabled. Therefore, it's crucial to have proper training and use all safety measures such as signage, hazard buffer zones, lockout, high-voltage PPE, tooling, and equipment.

Always assume that the vehicle's safety systems may fail, and that the vehicle could potentially cause harm. Don't let familiarity lead to complacency when it comes to health and safety during maintenance and repair work.

When working on hybrid and electric vehicles, safety is paramount due to the many risks involved, such as electrocution from high-voltage systems, strong magnetic fields, exposure to chemicals from damaged batteries, fire, and explosion. Always assess the risks and implement safety measures before starting any work.

- ☑ Only those with adequate training should work on high-voltage electrical systems.
- ☑ Always use the correct personal protective equipment (PPE) when working on these systems.
- ☑ High-voltage components, often identified by bright orange insulation and shielding, can cause severe injury or death if not handled correctly. Always use the right tools and equipment and check them before each use.
- ☑ If you're using electrical measuring equipment, ensure it is rated for the systems to be tested, accurate and calibrated.
- ☑ When replacing electrical or electronic components, make sure they meet the original equipment manufacturer (OEM) specifications. Using inferior parts or modifying the vehicle could void the warranty and affect vehicle performance and safety. Only replace electrical components if they comply with legal requirements for road use.

Throughout this book, you will find features that aim to support and enhance your understanding and use, such as:

The information in these boxes highlights safety features to consider when working on vehicles and electrical circuits, especially high-voltage systems. These features aim to minimise the risk of injury or damage to vehicles or equipment. Even if specific safety advice is not provided, always evaluate potential risks before starting any activity or diagnostic routine.

The guidance in these boxes is intended to support the information about the construction and operation of electric vehicle systems. It provides material that enhances understanding and strengthens knowledge of system components and testing methods.

Introduction

This feature explains the key terms related to electric vehicle operation, components, and diagnostic testing. Understanding and correctly using technical vocabulary is the foundation for effective repairs. Words highlighted in bold within the text are defined here.

These tips offer useful diagnostic advice for specific systems and components. Although not all of them may be relevant to your current task or vehicle, they may inspire ideas that you can modify and incorporate into your diagnostic routines. Always take care when implementing any diagnostic process to avoid the possibility of damage or injury to yourself, others, vehicles, or equipment.

The guidance in these boxes uses analogies to compare complex operational systems or component designs to simple concepts. The purpose of these explanations is to clarify and improve understanding, even though they are not scientifically accurate. However, they are only a tool to aid comprehension, not a replacement for correct information.

Preparing for assessment

The information in this book can help you with theory or practical assessments that measure your skills or competence in vehicle repairs or a recognised qualification. You may be able to use some of the evidence you produce for more than one qualification. You should make the best use of all your evidence to maximise the opportunities for cross-referencing between units or qualifications.

You should choose the type of evidence that suits the type of assessment you are undertaking (either theory or practical).

The types of evidence you could use are listed below:

- Direct observation by a qualified assessor
- Witness testimony
- Computer-based
- Audio recording
- Video recording
- Photographic recording
- Professional discussion
- Oral questioning
- Personal statement
- Competence/Skills tests
- Written tests
- Multiple-choice tests
- Assignments/Projects

Before taking a written or multiple-choice test, review the key terms related to the subject. Read all questions and answers thoroughly to understand what is being asked, as multiple-choice tests often have similar options that can be confusing.

For practical assessments, make sure you have had ample practice and feel confident in your ability to pass. Having a plan of action and a work method can be helpful.

Ensure you have the correct technical information, such as vehicle data, and the necessary tools and equipment. Check your work regularly to ensure accuracy and prevent issues from developing.

Always prioritise safety when performing any practical task.

Electrical Fundamentals

Chapter 1 Electrical Fundamentals

In this chapter, you will learn about the basic principles of electricity and how they apply to automotive systems. You will also learn about the common tools and measurements used to test, diagnose, and repair electrical problems in vehicles.

Electricity is the flow of electrons through a conductor, such as a wire or a metal. Electricity can power various vehicle systems, such as high-voltage hybrid and electric drive trains and low-voltage auxiliary circuits.

To understand how electricity works, you need to know some key concepts, such as voltage, current, resistance, and power.

To measure and test electrical circuits and components, you need to use some tools, such as multimeters, test lights, oscilloscopes, and scan tools. These tools can provide technicians with electrical system information and indicate whether a circuit has voltage or not.

Contents

- ❖ What is electricity **Page 6**
- ❖ Electrical units and terminology **Page 9**
- ❖ Ohms and power law **Page 14**
- ❖ Electric circuits **Page 15**
- ❖ Sources of portable power and circuit protection **Page 16**
- ❖ Electric and electronic components **Page 18**
- ❖ Electrical tooling and measurement devices **Page 23**

The automotive industry is a high-risk environment, especially when dealing with electrical systems. The hazards of electricity are well-known but can be easily ignored due to its invisible nature. This can lead to complacency if the fundamentals of electricity are not well understood. Even with this understanding, caution is necessary. Assume that any safety systems designed for protection have failed and take precautions to minimise the risk of injury or death. Always evaluate the risks associated with any activity and implement measures to eliminate or reduce the hazards involved in any task, diagnosis, or repair.

Additional risks associated with working on, or around electric electrical systems may include:

- ➤ Electrocution
- ➤ Strong magnetic fields
- ➤ Falling from height
- ➤ Short circuit
- ➤ Electrical discharge/arcing
- ➤ Fire and explosion
- ➤ Chemicals

Electrical Fundamentals

What is Electricity

The Discovery of Electricity

About 2500 years ago, a Greek scientist named Thales discovered that rubbing amber (fossilised tree sap) with a cloth attracted small dust and fluff particles. This was his discovery of static electricity. While Thales did not fully understand the phenomenon, he did document his findings.

Around 1550, William Gilbert, who was Queen Elizabeth I's doctor, discovered that rubbing a silk cloth on a glass rod could attract even heavier objects, like feathers. He called this phenomenon 'electricus', taking the name from the Greek word for amber, 'elektron', leading to the word electricity.

While static electricity is interesting, it's hard to convert into a usable energy source because electricity needs to move to be useful. In the late 18th century, two Italian scientists, Luigi Galvani, and Alessandro Volta, were competing with each other and ended up creating the first moving electricity, known as electric current. This electric **current** was produced through a chemical reaction and eventually led to the invention of the battery.

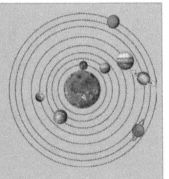

Understanding electricity can be challenging because it exists within tiny atoms.
You can visualise an atom as a mini solar system, with the **nucleus** acting as the sun and electrons orbiting around it like planets. The nucleus consists of positively charged protons and neutral neutrons, while the orbiting electrons carry a negative charge. When the electrons move from one atom to another, it is these electrons that generate electric current.

Atoms and Molecules

Every substance is composed of **molecules**, which are made up of **atoms**. For instance, water is a molecule denoted as H_2O, comprising two hydrogen (H) atoms and one oxygen (O) atom.

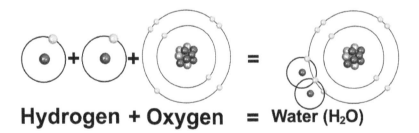

Figure 1.1 Hydrogen and Oxygen Making up a Water Molecule

Current - the flow of electric charge though a conductor.

Nucleus - the central part of an atom.

Atom - the smallest unit of matter that has the characteristic properties of a chemical element.

Molecule - a group of two or more atoms that are held together by chemical bonds.

Electrical Fundamentals

The number of **protons** and **electrons** varies among atoms, as depicted in the periodic table. This chart organises elements by atomic weight, which corresponds to the number of protons in their nucleus.

1 H																	2 He
3 Li	4 Be											5 B	6 C	7 N	8 O	9 F	10 Ne
11 Na	12 Mg											13 Al	14 Si	15 P	16 S	17 Cl	18 Ar
19 K	20 Ca	21 Sc	22 Ti	23 V	24 Cr	25 Mn	26 Fe	27 Co	28 Ni	29 Cu	30 Zn	31 Ga	132 Ge	33 As	34 Se	35 Br	36 Kr
37 Rb	38 Sr	39 Y	40 Zr	41 Nb	42 Mo	43 Tc	44 Ru	45 Rh	46 Pd	47 Ag	48 Cd	49 In	50 Sn	51 Sb	52 Te	53 I	54 Xe
55 Cs	56 Ba	57-71	72 Hf	73 Ta	74 W	75 Re	76 Os	77 Ir	78 Pt	79 Au	80 Hg	81 Tl	82 Pb	83 Bi	84 Po	85 At	86 Rn
87 Fr	88 Ra	89-103	104 Rf	105 Db	106 Sg	107 Bh	108 Hs	109 Mt	110 Ds	111 Rg	112 Cn	113 Nh	114 Fl	115 Mc	116 Lv	117 Ts	118 Og

		57 La	58 Ce	59 Pr	60 Nd	61 Pm	62 Sm	63 Eu	64 Gd	65 Tb	66 Dy	67 Ho	68 Er	69 Tm	70 Yb	71 Lu	
		89 Ac	90 Th	91 Pa	92 U	93 Np	94 Pu	95 Am	96 Cm	97 Bk	98 Cf	99 Es	100 Fm	101 Md	102 No	103 Lr	

Figure 1.2 The Periodic Table of Elements

Movement of Electrons

To generate an electric current, electrons need to move from one atom to another. Electron movement requires an external force or pressure, which can be created by magnetic fields or a chemical reaction.

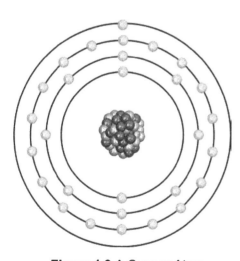

Electrons orbit the nucleus of an atom, similar to how planets orbit the sun due to gravity. In some atoms with simple structures, the attraction between the nucleus and electrons is very strong, making it difficult for electrons to move. Elements where electrons don't move easily are known as **insulators**.

Other atoms, however, have a weaker attraction force between the nucleus and electrons within their structure. Take a copper atom, for example, which has 29 electrons and 29 protons. The electrons orbit in increasingly larger circles. The outermost electrons, known as 'free electrons', have a weaker bond to the nucleus than those in simpler atoms. If an external pressure is applied, these free electrons can be made to move from one atom to another, creating an electric current. When electrons move easily, the element is known as a **conductor**.

Figure 1.3 A Copper Atom

In vehicles, we use conductors where we want electricity to flow easily, such as in wiring. We use insulators to restrict the movement of electricity, such as the coating on the outside of a high-voltage cable.

Certain elements, such as silicon and germanium, can be engineered into components that function as either conductors or insulators. They can even be switched between these two states, acting as controls for electronic systems. These versatile elements are known as **semiconductors**.

Electrical Fundamentals

For electrons to move from one atom to another, they need a continuous path, known as a **circuit**. This allows an electron to be replaced by another one from behind as it moves. Without a complete circuit, electrons can't flow because the last electron in the conductor has nowhere to go. If the circuit is interrupted, it loses **continuity**.

> **Proton** - a subatomic particle that has a positive electric charge and is found in the nucleus of every atom.
>
> **Electron** - a subatomic particle that has a negative electric charge and is one of the main components of matter.
>
> **Insulator** - an electrical component that restricts or prevents the flow of electric current.
>
> **Conductor** - an electrical component which allows the flow of electric current.
>
> **Semiconductor** - a material that can have the properties of both a conductor or an insulator when used in an electric circuit.
>
> **Circuit** - a continuous, unbroken loop that allows the steady flow of electricity.
>
> **Continuity** - refers to a complete, unbroken conductor that enables the uninterrupted flow of electricity.

Electromagnetism

Electricity and magnetism are closely related, like two sides of the same coin. Both have positive and negative poles, or north and south, and both can attract and repel.

When a magnet passes a copper conductor (wire), the magnetic attraction moves electrons through the conductor, creating an electric current. Conversely, when an electric current passes through a copper conductor, it generates an invisible magnetic field. The magnetic effect of an electric current can cause movement through attraction or repulsion. This movement can be harnessed to create a motor.

Similarly, the movement of magnets past a conductor can generate an electric current, which is the principle behind a generator.

> ➤ Motors convert electrical energy into mechanical energy.
> ➤ Generators convert mechanical energy into electrical energy.

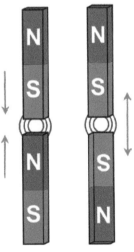

Figure 1.4 Magnets Attracting and Repelling

Chemical Reactions

Electrical energy can be converted into chemical energy, which can be stored and transported in a battery. This process is reversible, meaning chemical energy can be turned back into electricity. Therefore, a charged battery provides a portable source of electricity that can be used as required.

Figure 1.5 A Battery

Electrical Fundamentals

Electrical Units and Terminology

To better comprehend how electricity functions in a circuit, we use specific units of measurement.

It is important to use electrical terminology correctly to ensure misinterpretation does not promote incorrect testing and diagnosis.

The following **Table 1.1** presents the four primary units associated with electricity.

Table 1.1 Electrical units	
Volts 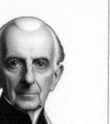	Voltage, named after Alessandro Volta, is the electrical pressure or potential force in any part of an electrical circuit. There are two main types of voltage in electrical circuits: 1. **Electromotive Force (EMF)**: This is the potential pressure when all electrical devices are turned off and no current is flowing, often considered as the **open circuit voltage (OCV)**. 2. **Potential Difference (Pd)**: This is the voltage drop caused by electricity flow when the circuit is active or switched on. When the circuit is active, this is known as the **closed-circuit voltage (CCV)**. Voltage is often represented in technical information or documentation as: V - Volts named after Alessandro Volta. E - EMF to describe electromotive force.
Amps 	Amps, named after André-Marie Ampère, are the units used to quantify the amount of electricity in any part of an electrical circuit. It's measured when electricity is allowed to flow in a circuit, a phenomenon known as current. There are two main types of electrical current: 1. Direct Current (DC): This is electricity that flows in one direction only. 2. Alternating Current (AC): This is electricity that oscillates back and forth in a circuit. Regardless of where you measure it in the circuit (at the beginning, middle, or end), the amperage remains the same. Current (amps) is often represented in technical information or documentation as: I - which originates from the French phrase intensité du courant, (current intensity). A - amps, named after André-Marie Ampère.

Electrical Fundamentals

Table 1.1 Electrical units	
Ohms 	Ohms, named after Georg Ohm, are the units used to measure resistance to electrical flow. Resistance directly impacts the functioning of any electrical circuit as it attempts to slow down the flow of electricity. As resistance increases in a circuit, both current and voltage decrease. This can limit the operation of electrical components. While resistance can be used to control electrical components in some circuits, high resistance is generally undesirable. Resistance in electrical circuits is often closely associated with heat. It could be said that heat is an indication of resistance, and vice versa. Resistance (ohms) is often represented in technical information or documentation as: Ω - the Greek letter Omega (meaning 'great'), which sounds similar to ohms and ensures that the letter 'O' is not confused with a zero.
Watts 	Watts, named after James Watt, are the units used to measure electrical power produced or consumed. Power is essentially the speed at which work is done. In the context of electrical components, a higher wattage indicates a more powerful component that uses more electrical energy. Power is often represented in technical information or documentation as: W - named after James Watt. P - to represent the word power. The word horsepower is often attributed to James Watt but considered to be an imperial measurement. The Si unit of power is the Watt. 1 horsepower (hp) is equivalent to approximately 746 Watts. 1.34 horsepower (hp) equals 1 Kilowatt (KW).

Electromotive force (EMF) - a measure of electrical potential in volts. It represents the maximum voltage available at the time of measurement. Often associated with an open circuit, it's the peak voltage waiting to be used when connected to a power source like a battery, or when generated. In simpler terms, EMF is like the highest pressure of electricity ready to do work.

Potential Difference (Pd) - also known as voltage, is measured when a circuit is active, and current is flowing. This value can be higher or lower than the initial potential, depending on the direction of the current flow. If the circuit is complete, a potential difference will cause the current to flow. However, the amount of current is proportional to the resistance in the circuit. In simpler terms, Potential Difference is like the force that pushes the current through a circuit when it's switched on.

Electrical Fundamentals

Open circuit voltage (OCV) - is the voltage measured when the circuit is turned off. It's often associated with the total potential of the circuit. In simpler terms, OCV is like the stored energy in a switched-off circuit.

Closed-circuit voltage (CCV) - is the voltage measured when the circuit is turned on. It's often associated with the potential difference in the circuit. In simpler terms, CCV is like the active energy in a switched-on circuit.

Regarding the hazards of voltage and current (amps), voltage is often considered the dangerous element, which is why warning signs typically state 'Danger High Voltage'. If a voltage exceeds the touch threshold of dry human skin, it can cause an electric current to flow, and it's this current that can cause harm. Keeping the voltage potential low reduces the risk of electric shock or electrocution. However, even with low voltage, there's still a risk of a short circuit that can lead to arcing, fire, or explosion.

Consider lightly placing your hand on a sharp, upturned nail. Your skin has resistance, so as long as the pressure is light, the nail won't cause damage. However, if you increase the pressure on your hand, the nail will eventually overcome the skin's resistance and pierce it.
In this analogy, the pressure is like voltage. It's the cause of any injury. But the size of the nail, which is similar to the amount of current, determines the extent of the damage. If the pressure or voltage stays below the threshold where it can overcome skin resistance, then the size of the nail or current doesn't matter. That's why warning signs say, 'Danger High Voltage' and not 'Danger High Current'.

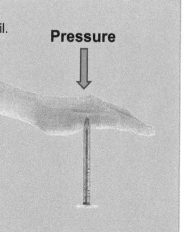

AC and DC - Three Phase

Three Phase Electric current comes in two forms:
- Direct Current (DC): This is when electrons in a circuit move in only one direction, driven by a stable potential. In simpler terms, DC is like a one-way street for electrons.
- Alternating Current (AC): This is when electrons change direction periodically, driven by an oscillating potential. An AC output is often called a 'phase'. To put it simply, AC is like a two-way street where electrons can go back and forth.

To boost the amount of electrical energy in an AC system, you can combine multiple AC phases. The most common method is to use three separate AC phases that are spaced 120 degrees apart. This is known as 'Three Phase'. So, it's like having three two-way streets for electrons to move along, which can carry more electric traffic.

Electrical Fundamentals

 Alternating current (AC) voltages are often described using their RMS values. RMS stands for root-mean-square, and it is a way of measuring the effective value of an alternating current (AC) or voltage. It is equal to the direct current (DC) or voltage that would produce the same amount of heat or power in a resistor. It roughly equates to the peak value of an AC wave divided by the square root of 2 and is proportional to a DC voltage equivalent.

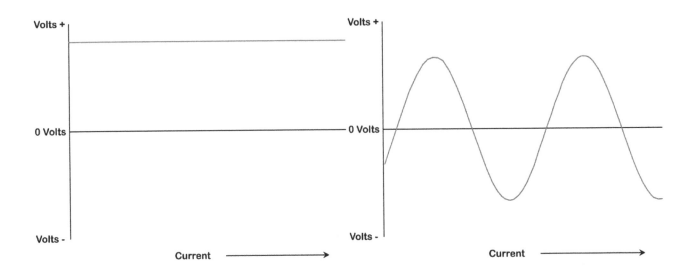

Figure 1.6 Direct Current (DC)

Figure 1.7 Alternating Current (AC)

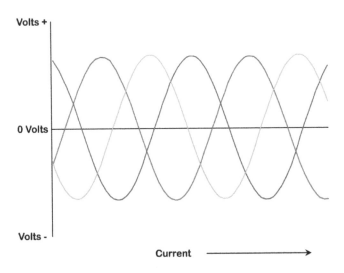

Figure 1.8 Three Phase Current

Electrical Fundamentals

Understanding a direct current circuit can be tricky, so let's use a water tower analogy to explain it:

> Reservoir of Water at the Top (Battery): This is like the battery in a circuit. It stores the energy, just like the reservoir stores water.

> Pipe Leading from the Bottom of the Reservoir (Wire): This is like the wire in a circuit. It's the path that the energy (or water) travels along.

> Tap on the End of the Pipe (Switch): This is like a switch in a circuit. When you open the tap, water flows; when you close it, the water stops. Similarly, a switch controls whether electricity flows in a circuit.

> Water Wheel at the End (Motor): This is like a motor in a circuit. The flowing water turns the wheel, just as flowing electricity powers the motor.

Therefore, in this analogy, when you open the tap (switch), water (electricity) flows from the reservoir (battery) through the pipe (wire) and turns the wheel (motor). When you close the tap, everything stops.

> Quantity of Water (Current): The amount of water flowing through the pipe is like the current, or amperage, in an electric circuit.

> Water Pressure (Voltage): The pressure of the water in the pipe is similar to the voltage in an electric circuit.

> Tap Slowing Water Flow (Resistance): The way a tap slows down the flow of water is like resistance in an electric circuit.

> Water Wheel's Work Rate (Power): The speed at which the water wheel works is like the power in an electric circuit.

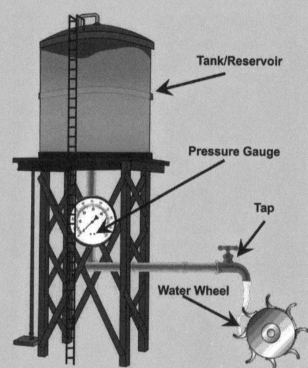

Therefore, in this analogy, the amount and pressure of water represent the current and voltage. The tap represents resistance, and how fast the water wheel works represents power.

Electrical Fundamentals

Ohms and Power Law

Ohms Law

Ohm's Law explains how voltage, current (amps), and resistance (ohms) in a circuit are related. If you change one of these factors, it affects the others. Here's a simpler explanation using the water analogy:

1. Voltage (Pressure): If you increase the voltage in a circuit, it's like increasing the water pressure in a pipe. This makes more current flow, just like more water would flow through the pipe.

2. Resistance (Tap): If you increase the resistance in a circuit, it's like partially closing the tap on a pipe. This makes less current flow, just like less water would flow through the pipe.

Georg Ohm explained this with these formulas:
- Current (Amps) = Voltage ÷ Resistance
- Resistance (Ohms) = Voltage ÷ Current
- Voltage = Current × Resistance

Therefore, with Ohm's Law, if you know two of these measurements, you can calculate the third one. The Ohm's Law triangle is a handy tool for doing these calculations.

Figure 1.9 Ohms Law Triangle

In electrical equations, we often use the following symbols:
- **V** represents volts.
  Sometimes, you might see the letter 'E' used instead to represent EMF (Electromotive Force), but it still means volts.
- **I** stands for amps.
  This letter is used to represent intensité du courant, (current intensity).
- **R** is used for ohms.
  We use 'R' for resistance to avoid confusion with zero.

The Ohm's Law triangle can help you calculate unknown units.

Here's how to use it:
Cover the unknown unit with your thumb, and the remaining letters form the calculation you need.
For example, if you don't know the amperage (I), cover the 'I' in the triangle. You'll be left with V ÷ R, which means volts divided by resistance.

Ohm's Law, which explains the relationship between voltage, resistance, and currant (amperage), can be a useful tool for diagnosing faults in an electrical circuit. By taking measurements and comparing them using Ohm's Law calculations, you can identify where the fault might be:
- Voltage (Pressure): If the voltage is lower than expected, the performance of the component might be reduced. If it's higher than expected, it could cause the component to be overworked and damaged.
- Current (Quantity/Amps): If the current is lower than expected, the component might not operate correctly. If it's higher than expected, it could mean that the component or system is being overworked.
- Resistance (Ohms): If the resistance is lower than expected, it could indicate a short circuit, where current is taking an alternative path to earth. If it's higher than expected, it could consume electrical energy and reduce system performance.

Therefore, if your voltage, current, or resistance measurements are different from what you expect, it could indicate a problem with your circuit.

Electrical Fundamentals

Power Law

Power, measured in watts, can be calculated similarly to Ohm's Law:
> Current (Amps) = Power ÷ Voltage
> Voltage = Power ÷ Current
> Power (Watts) = Current × Voltage

You can use a power triangle, like the Ohm's Law triangle, to help with these calculations.

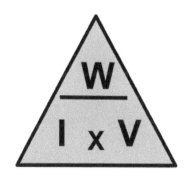

Figure 1.10 Power Law Triangle

> **W** represents power in watts.
> Sometimes, you might see the letter 'P' used instead to represent power, but it still means watts.
> **V** represents volts.
> Sometimes, you might see the letter 'E' used instead to represent EMF (electromotive Force), but it still means volts.
> **I** stands for amps.
> This letter is used to represent intensité du courant, (current intensity).

The Power Law triangle can help you calculate unknown units.

Here's how to use it:
Cover the unknown unit with your thumb, and the remaining letters form the calculation you need.
For example, if you don't know the amperage (I), cover the 'I' in the triangle. You'll be left with W ÷ V, which means watts divided by volts.

Electric Circuits

For electrons to move from one atom to another, they need a continuous path, or a **circuit**. This allows an electron to move forward as it's replaced by another one from behind. If there's no circuit, the electrons can't flow because the last electron in the conductor has nowhere to go. If the circuit is broken, we say it has lost continuity.

In other words, electrons need a complete loop to move. If the loop is broken, the electrons can't move because they have nowhere to go.

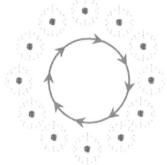

Figure 1.11 Copper Atoms Forming a Loop

Series and Parallel Circuits

Two main types of electrical circuit are used in the construction of motor vehicles:
> **Series**
> **Parallel**

Series Circuit

In a series circuit, devices are connected one after another in a single line. They all share the same circuit, so they divide the electricity based on how much power each device uses. If you add more devices to the circuit, each one gets only a portion of the available voltage. The more power a device needs, the more electricity it uses.

In simpler terms, in a series circuit, all devices are lined up in a row. They share the electricity, and each device only gets a part of it. The stronger the device, the more electricity it takes.

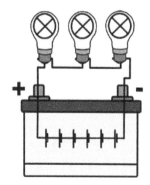

Figure 1.12 Bulbs Connected in Series

In a series circuit, if any one device stops working, it breaks the circuit and stops the flow of electricity. This means all the other devices in the circuit will also stop working.

Electrical Fundamentals

Parallel Circuit

In a parallel circuit, devices are connected side-by-side. Each device has its own power supply and return path to the supply. Because of this, all devices receive the full voltage and can operate at full power.

When you add a device to a parallel circuit, it creates a new branch or pathway. This allows more current to flow, which is inversely proportional to the resistance in that branch. This is known as a potential divider.

If one device in a parallel circuit stops working, the others will continue to work. This is because each device has its own separate pathway for electricity.

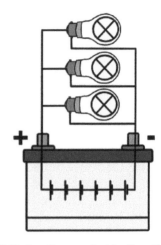

Figure 1.13 Bulbs Connected in Parallel

In a series circuit, the components are connected one after the other and the current flows through each component in turn. The voltage across each component is different and is determined by its resistance according to Ohm's law. The total resistance of a series circuit is equal to the sum of the individual resistances.

In a parallel circuit, the components are connected on different branches of the circuit and the voltage across each component is the same. The current flowing through each branch is determined by its resistance according to Ohm's law. The total resistance of a parallel circuit is less than any of the individual resistances.

Circuit - a continuous and unbroken loop.

Series - connected one after another.

Parallel - connected side-by-side.

Sources of Portable Power

Electric vehicles can use various power sources to make or store energy. Here are some examples:

Batteries: Batteries are electrical storage devices that use chemistry to create a potential difference between a positive and negative **electrode**. When connected in a circuit to a device, they provide a portable source of power. In simpler terms, batteries are like little chemical factories that produce electricity.

Figure 1.14 Battery Symbol

Generators: Generators can be thought of as mechanical electricity pumps. They use moving or rotating magnets to induce current in corresponding wire coils. They need a source of mechanical energy to generate electric current. So, you can think of generators as machines that turn physical movement into electricity.

Figure 1.15 Generator Symbol

Electrical Fundamentals

Capacitors: Capacitors are temporary storage devices for electrical energy. Unlike a battery, which uses chemistry to create an electrical potential, a **capacitor** can't generate its own electricity and must be charged from an external source. You could compare a capacitor to a bucket that can be filled and emptied with electrical energy.

Figure 1.16 Capacitor Symbol

Circuit Protection

Electricity naturally seeks the easiest route back to its source, taking the path of least resistance. If it finds an alternative path that bypasses a device (consumer), the resulting electric current can generate excessive heat, potentially damaging components or causing fires or explosions. To mitigate this risk, vehicles often incorporate circuit protection devices, including:

Fuses: A fuse is a weak link in the circuit, often a thin wire with a current rating slightly above the intended current flow in the system. It's designed to burn out if there's a rapid increase in current, preventing further damage to the circuit. Once a fuse burns out, it creates an open circuit and stops the current flow, potentially saving more expensive parts.

Figure 1.17 Fuse Symbol

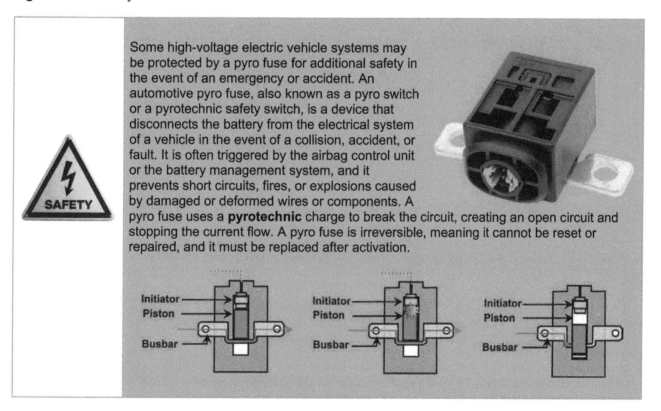

Some high-voltage electric vehicle systems may be protected by a pyro fuse for additional safety in the event of an emergency or accident. An automotive pyro fuse, also known as a pyro switch or a pyrotechnic safety switch, is a device that disconnects the battery from the electrical system of a vehicle in the event of a collision, accident, or fault. It is often triggered by the airbag control unit or the battery management system, and it prevents short circuits, fires, or explosions caused by damaged or deformed wires or components. A pyro fuse uses a **pyrotechnic** charge to break the circuit, creating an open circuit and stopping the current flow. A pyro fuse is irreversible, meaning it cannot be reset or repaired, and it must be replaced after activation.

Circuit Breakers: A circuit breaker works similarly to a fuse. It's connected in series with the circuit it's protecting. If there's excess current, it generates heat. But instead of burning out like a fuse, a circuit breaker acts like a heat-sensitive switch that opens the circuit and stops the current flow. Unlike fuses, which need to be replaced once triggered, circuit breakers can often be reset and reused.

Figure 1.18 Automotive Circuit Breaker Symbol

Electrical Fundamentals

Residual Current Devices (RCDs): An RCD is a safety device that monitors any imbalance between the incoming and outgoing current in a circuit. It uses a differential current transformer to measure any difference in circuit current and can cut off the supply to help prevent electric shock if triggered. RCDs are most commonly associated with circuits related to charging an electric vehicle from a mains supply.

Electrode - an electrical conductor that makes contact with a non-metallic part of a circuit, such as an electrolyte.

Capacitor - an electronic device that stores electrical energy in an electric field by accumulating electric charges on two closely spaced surfaces that are insulated from each other.

Pyrotechnic - a small explosive charge. Pyro - comes from Greek, where it has the meaning fire, heat, or high temperature.

Electric and Electronic Components

Motors

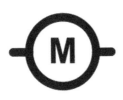

Electric motors are devices that convert electrical energy into mechanical energy. They work using the principle of electromagnetism, which is the interaction between electric currents and magnetic fields. Electric motors have two main parts: a stationary component and a moving component, which, when acted on by an electromagnetic force, produces mechanical movement. There are two main types of electric motor: those that rotate and those that act with a linear motion. A linear motor will often be categorised as a solenoid. Depending on the design, electric motors can be operated by using direct current (DC) or alternating current (AC).

Figure 1.19 Motor Symbol

Relays

A relay is a component that uses an electromagnet to switch a high-current circuit on and off with a low-current signal. It contains an electromagnetic coil and a spring-loaded switch.
The coil is a wire wrapped around a metal core that creates a magnetic field when current flows through it. The switch is a set of movable contacts, which are the terminals that connect or disconnect the high-current circuit. Spring tension holds the switch in its normal position when the electromagnetic coil is not energised.

When the coil receives a low-current feed from a control device, such as a switch or a sensor, it generates a magnetic force that attracts the spring-loaded contacts and moves them towards the coil. This movement causes the contacts to close or open, depending on the type of relay. A normally open relay closes the contacts when the coil is energised, while a normally closed relay opens the contacts when the coil is energised. This allows the high-current circuit to be turned on or off by the low-current signal, without any direct contact between them. This protects the control device from damage, reduces voltage drop and power loss in the circuit. When the coil loses the low-current signal, the magnetic field collapses and the spring pushes the contacts back to their original position.

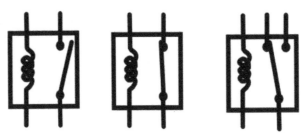

Figure 1.20 Relay Symbols

Electrical Fundamentals

Inductors

An inductor is an electrical component that stores energy in a magnetic field when an electric current flows through it. An inductor is often made from insulated wire wound into a coil, which can have different shapes and sizes depending on the application.

An inductor can also step-up voltage, as when it is switched off, the current stops flowing, and the resulting collapse of the magnetic field induces a back **electromotive force (EMF)**. This is a voltage spike with a reverse polarity.

Figure 1.21 Inductor Symbol

Transformers

An electrical transformer is a component that transfers electrical power from one circuit to another. It uses the principle of electromagnetic **induction** and mutual induction, which means that a changing current in one coil induces a voltage in another coil that is magnetically linked to it.

A transformer will normally contain:

- A core made from laminated iron or silicon steel to provide a low **reluctance** path for the **magnetic flux**.
- A primary winding, which is connected to the input voltage source and produces the magnetic field when current flows through it.
- A secondary winding, which is connected to the output load and receives the induced voltage from the magnetic field.

When the primary winding is connected to an alternating voltage source, an alternating current flows through it and generates an alternating magnetic field in its core. This magnetic field passes across the secondary winding and induces an alternating voltage in it through induction.

The size of the induced voltage in the secondary winding depends on the number of turns of wire in the primary and secondary coils. The turns ratio determines whether the transformer is a step-up or a step-down transformer.

Figure 1.22 Transformer Symbol

- A step-up transformer has more turns in the secondary winding than the primary winding and increases the output voltage.
- A step-down transformer has fewer turns in the secondary winding than the primary winding and decreases the output voltage.

Electromotive force (EMF) - a measure of electrical potential in volts. It represents the maximum voltage available at the time of measurement.

Induction - the process by which an electrical conductor becomes electrified when near a charged body, or a magnetisable body becomes magnetised when in a magnetic field.

Reluctance - a measure of the opposition to magnetic flux, which is the flow of magnetic field lines through a material. It is similar to electric resistance in a circuit, which measures the opposition to electric current.

Magnetic Flux - a measure of the amount of magnetic field passing through a surface.

Electrical Fundamentals

Resistors

Electrical resistors are components that provide resistance to the flow of electric current in a circuit. Resistors can be used for various purposes, such as limiting the current, dividing the voltage, and biasing active circuit elements.

There are several types of resistors, including fixed resistors and variable resistors.

Fixed resistors have a constant resistance value that does not change with temperature, time, or voltage.

Some examples of variable resistors are shown below:

- Thermistors - resistors that change their resistance with temperature.
- Potentiometers - a form of variable resistor that can vary their resistance value by adjusting a knob, slider, or wiper.
- Rheostats - a form of variable resistor used to control the current in a circuit.

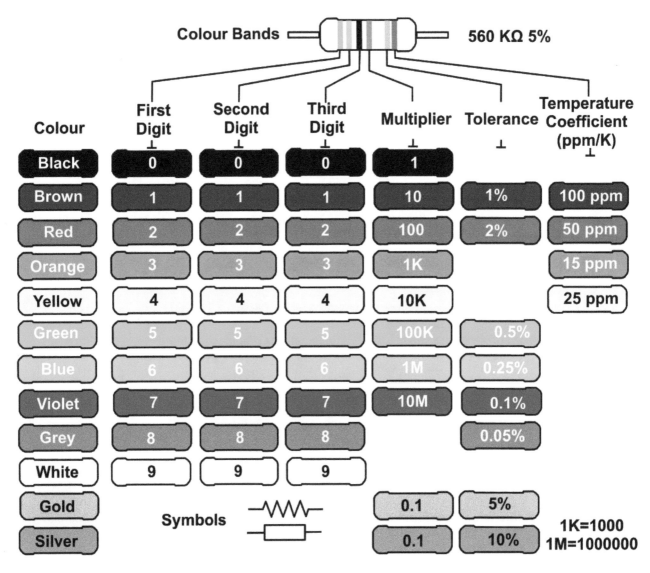

Figure 1.23 Fixed Resistor Colour Codes

Electrical Fundamentals

Diodes

A diode is an electronic component that only allows electric current to flow in one direction. It is made of two types of semiconductor material, called P-type (positive) and N-type (negative), that are joined together to form a PN junction. The P-type material is engineered to have fewer electrons than the N-type and is often described as having 'holes'. The PN junction creates a barrier known as a depletion layer, which prevents an easy flow of electrons between the two semiconductor elements.

When a diode is connected to a voltage source, the potential difference can push electrons from the negative side to the positive side across the depletion layer, as there is space (holes) for the electrons to fill, and current flows. However, if the polarity is reversed, the negative side is full, and the holes cannot be pushed across the depletion layer.

Figure 1.24 Diode Symbol

Diodes can be damaged by over-voltage. A voltage potential higher than the design capability with reverse polarity will cause an issue known as the breakdown voltage. At this point, the diode's ability to prevent current flow fails and allows a large current to flow in the reverse direction. This can damage the diode if not controlled.

There are several types of diodes that have different characteristics and applications. Some of the common types are:

- Zener diode - A diode that is designed to break down at a set voltage value and allow current flow in the reverse direction.
- Light-emitting diode (LED) - A diode that emits light when connected to a circuit in one direction only.
- Photodiode - A diode that generates current when exposed to light.
- Schottky diode - A diode that has a metal-semiconductor junction instead of a PN junction and has a lower forward voltage drop and faster switching speed.
- Tunnel diode - A diode that has a very thin PN junction and exhibits negative resistance in the forward direction, meaning that the current decreases as the voltage increases.

Transistors

Transistors are semiconductor electronic components that can control the flow of electrical current in a circuit. They can act as either amplifiers or switches with no moving parts.

A transistor controls the flow of electric current. Transistors use semiconductor materials composed of P-type and N-type materials in a similar manner to diodes. They are arranged so that they generally have three parts, which are joined to create either PNP (positive-negative-positive) junctions or NPN (negative-positive-negative) junctions. These arrangements will either block the flow of current or allow the flow of current depending on their design. A transistor will be joined to a circuit with an input current connection, often known as the collector or source, an output current connection, often known as the emitter or drain, and a voltage connection in the middle semiconductor component, often called the base or gate.

By switching a small voltage on and off, the base/gate section can be changed between two states:

- A conductor, which will allow current to flow between the collector/source and the emitter/drain.
- An insulator, which prevents the flow of current between the collector/source and the emitter/drain.

In this way, a small voltage can be used to determine or regulate the flow of current in an electrical circuit, allowing it to act as a switch or amplifier.

Electrical Fundamentals

A good way of imagining a transistor's operation is to think of it like a farmer opening and closing a gate to allow sheep into a field.
On one side of the gate, the sheep are waiting to enter the field; this corresponds to the collector or source.
On the other side of the gate is the field, where the sheep will graze; this corresponds to the emitter or drain.
The farmer can control the flow of sheep into the field by opening and closing the gate, which is similar to applying a voltage at the transistor's base or gate.

There are two main types of transistors: bipolar junction transistors (BJTs) and metal-oxide-semiconductor field-effect transistors (MOSFETs).

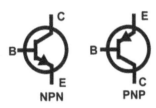

> A BJT has three terminals: the base, the collector, and the emitter. The base is the input terminal that controls the current flow between the collector and the emitter.
> A MOSFET has four terminals: the gate, the source, the drain, and the body. The gate is the input terminal that controls the current flow between the source and the drain. The source is the terminal that supplies the current. The drain is the terminal that receives the current. The body is the terminal that connects to the reactive section of the semiconductor.

Figure 1.25 Transistor Symbols

Insulated-Gate Bipolar Transistors (IGBT)

An insulated-gate bipolar transistor (IGBT) is a type of power transistor that combines the features of both MOSFETs and bipolar transistors. They can switch high currents and voltages with great efficiency and speed and are used in the inverter units of electric vehicles to convert the direct current (DC) from the high-voltage battery to alternating current (AC) for the drive motors.

Hall Effect

Discovered by Edwin Hall in 1879, the Hall effect is a phenomenon that occurs when a current-carrying conductor is placed at right angles to a magnetic field. Due to a process known as the Lorentz force, the electrons in the conductor are deflected to one side as they pass the magnetic field. If a voltage measurement is taken at this point of deflection across the conductor, a potential difference (voltage) can be measured, which is proportional to the strength of the magnetic field and the amount of current flowing. The potential difference from the Hall effect can be used to calculate current flow or magnetic strength. Therefore, the Hall effect can be used in different types of sensors that need to measure current or position. Some Hall effect sensors contain an analogue-to-digital converter, meaning that a square waveform digital output can be observed on some sensor types.

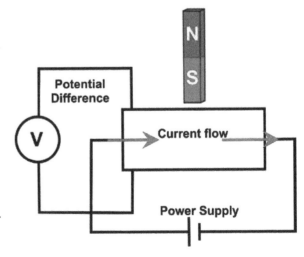

Figure 1.26 Hall Effect

Electrical Fundamentals

Electrical Tooling and Measurement Devices

Multimeters

A multimeter is an electrical testing tool designed to measure various units within an electrical circuit. There are two types of multimeters: one with a needle that moves over a scale (analogue), and one with a screen that shows numbers (digital).

To use a multimeter on a high-voltage system, it needs to have a high-voltage rating. This rating is indicated by the equipment category, which also shows the level of insulation that protects the operator. The higher the category, the better the protection.

For use on an electric vehicle's high-voltage system, the multimeter must be at least CAT III (CAT 3) 1000 volts. The voltage rating should always be higher than the expected system voltage.

Any test leads and probes should also match the multimeters category rating. Always check them for damage before use and test the meter on a known good source to prove correct operation and accuracy.

Analogue Multimeters

Analogue multimeters are devices that use a moving needle across a graduated scale to measure electrical readings in a circuit. They were traditionally known as 'AVO meters', which stands for amps, volts, and ohms.

The accuracy of analogue meters depends on the user's ability to read the scale correctly, which can be challenging. The needle's position between two units could represent any fraction, depending on the scale provided by the manufacturer. Also, analogue multimeters have an upper range limit. If the needle moves all the way to the end of this scale, this is called full-scale deflection (FSD).

Figure 1.27 An Analogue Multimeter

Digital Multimeters

Digital multimeters display measurements as numbers on a liquid crystal display (LCD) screen, making them easy to read accurately.

There are two common types of digital multimeters:

> **Manual Multimeters**: With a manual multimeter, the user selects the unit and the scale to be measured, usually by turning a dial on the front of the multimeter.

> **Autoranging Multimeters**: With an autoranging multimeter, the user selects the unit, but the multimeter automatically selects the scale of that unit. When using an autoranging multimeter, it's important to ensure your reading is accurate by noting the scale of the unit being displayed.

For example, if measuring voltage, the scale could be in millivolts, volts, kilovolts, or megavolts.

Figure 1.28 A Digital Multimeter

Electrical Fundamentals

Using a Digital Multimeter

A digital multimeter is a versatile tool that can measure various electrical units, including volts, amps, and ohms. Some models also offer additional features like temperature measurement, frequency measurement, diode testing, transistor tests, and audible continuity testing.

The units of volts and amps can be further categorised into Direct Current (DC) and Alternating Current (AC).

On a multimeter:

- The DC scale is usually represented by a straight line with several dots underneath it . This unique symbol helps avoid confusion with a minus sign (if only one line was used) or an equals sign (if two lines were used).
- The AC scale is typically represented by a wavy line, symbolising the alternating nature of the current ∾.
- The ohms scale is represented by the Greek letter omega (Ω) to avoid confusion with zero (if 'O' was used).

Using a Multimeter to Check Voltage

A multimeter can be used as a voltmeter to measure the voltage difference in an electric circuit. Here's how:

Figure 1.29 Using a Voltmeter

1
- Connect the probes to the multimeter:
- The black lead goes into the common socket.
- The red lead goes into the voltage socket.

2
- For **low-voltage** systems in vehicles, which typically use direct current (DC), select the scale with the straight and dotted lines.

3
- **High-voltage** systems use both alternating current (AC) for charging and drive systems, and DC at the battery and capacitors. Make sure to select the correct type and scale for the circuit you're testing.

4
- Connect the voltmeter in parallel across the circuit, following any **high-voltage** warning instructions and wearing PPE if necessary.

5
- For a **low-voltage** circuit, which often uses the vehicle frame or chassis as the negative return to the battery, connect the black lead to a good earth source like the auxiliary battery negative terminal, metal bodywork, or engine (if hybrid).

6
- Use the red lead to probe the electrical circuit being tested.

7
- A **high-voltage** circuit uses a fully insulated return, so both red and black probes must be placed in parallel across the positive and negative of the circuit being tested. Remember to wear high-voltage personal protective equipment (PPE).

Electrical Fundamentals

Using a voltmeter on a high-voltage vehicle circuit can be risky. When you connect a multimeter, particularly for voltage testing, you create a parallel path across the circuit with the multimeter probes. This often requires you to hold the test probes, creating another parallel connection with your hands. A voltmeter operates by having a large internal resistance to limit current flow and protect the circuits while measuring voltage. Electricity tends to follow the path of least resistance. If the leads or probes are damaged, this could create a parallel connection with your hands, causing current to pass through them and across your chest area. This could result in injury or even death. Therefore, when using a multimeter on a high-voltage circuit, high-voltage personal protective equipment (PPE) must always be worn.

Using a Multimeter to check Electrical Resistance

A multimeter can be used as an ohmmeter to test electrical resistance. Here's how:

1
- Always ensure the power is off and the component to be tested is disconnected from the circuit before checking for electrical resistance.

2
- Connect the probes to the multimeter:
- The black lead goes into the common socket.
- The red lead goes into the socket marked with the omega symbol (Ω).

3
- Calibrate the ohmmeter for accuracy before taking any measurements.
- Set the selector dial to the lowest ohms setting and touch the probe tips together.
- The readout should show zero or near zero.
- If any figures appear on the screen, they will need to be subtracted from your final results.

4
- When the leads are disconnected, you should see OL (off limits) or the number 1, which represents infinity.

5
- Connect the ohmmeter in parallel across the component to test the electrical resistance.

You can also use the ohmmeter to check for continuity, for example:

To check a wire for continuity, place the red and black probes at each end of the wire. The screen should display a very low resistance reading.

To check a switch's operation, connect the probes across the terminals and operate the switch.
- In the off position, the display should read OL (off limits) or infinity.
- In the on position, it should be close to zero.

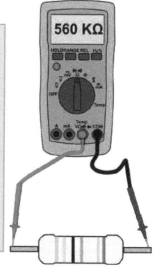

Figure 1.30 Using an Ohmmeter

Electrical Fundamentals

An ohmmeter uses its internal battery to generate voltage, which causes a current to flow through the component being tested. It then applies Ohm's Law to calculate and display the resistance on the screen. Therefore, any power in the circuit or component being tested must be removed. Any voltage differing from the meter's internal battery can lead to incorrect calculations, potential damage to the component or meter, and even risk to the operator.

Using an ohmmeter to check for high resistance in an electrical circuit can be misleading. For instance, a 12-volt lighting circuit drawing 10 amps with a bad connection causing a 0.1 Ω resistance will reduce the overall circuit voltage by 1 volt (Ohm's Law: 0.1 Ω x 10 A = 1 V). However, the same 0.1 Ω resistance in a motor circuit drawing 100 amps will reduce the overall circuit voltage by 10 volts (Ohm's Law: 0.1 Ω x 100 A = 10 V). This low resistance may have little or no effect on the lighting circuit but could cause complete failure of the motor circuit.

Therefore, it's better to use a volt drop test when checking for high resistance as it provides a clearer indication of why a circuit may not be working correctly.

Using a Multimeter to Measure Electrical Current (Not Suitable for High-Voltage Systems)

A multimeter can be used as an ammeter to measure the electrical current in a circuit.

Figure 1.31 Using an Ammeter

Here's how:

1. Be careful when using an ammeter as incorrect connection can damage the multimeter.

2. Connect the probes to the multimeter:
   - The black lead goes into the common socket.
   - The red lead goes into the socket used for measuring amps (usually separate from the one used for volts or ohms).

3. Turn the selector dial to the amps measurement.

4. Ensure the circuit being tested is switched off.

5. Break into the circuit being tested, taking care to avoid short circuits.

6. Connect the ammeter in series, turn on the circuit, and measure the current.

Electrical Fundamentals

A good place to connect an ammeter is at the fuse box – remove the fuse completely and replace it with the ammeter connected in series.

Never use an ammeter on high-voltage systems as it can cause damage, injury, or even death due to the nature of voltage and current in these systems.

Always remember not to connect an ammeter in parallel across a circuit. An ammeter typically has a very low internal resistance, so connecting it in parallel creates a short circuit. This leads to excessive current flow and can damage the ammeter. Also, keep in mind that the amount of current you can measure may be limited to around 10 amps, depending on the quality of the ammeter.

Additional Multimeter Functions

Many multimeters offer more than just voltage, amperage, and resistance checks. They come with extra features, some of which are described in the next section.

Audible Continuity Testing

Some multimeters have an audible continuity tester, which lets you check the continuity of an electrical component without looking at the screen.

Here's how:

1
- Connect the probes to the multimeter:
- The black lead goes into the common socket.
- The red lead goes into the socket marked with the omega symbol (Ω).

2
- Turn the dial to the audible continuity test setting.

3
- To calibrate the meter and ensure it's working correctly, touch the probes together; you should hear a tone.

4
- Like resistance testing, turn off the circuit power and disconnect the component being checked.

5
- Connect the red and black probes to the conductor's terminals.
- If continuity exists, you'll hear a tone.

Electrical Fundamentals

Diode Testing

Most multimeters have a diode test feature. A diode acts like a one-way valve for electricity. The test is conducted similarly to the continuity test.

Here's how:

1.
   - Connect the probes to the multimeter:
   - The black lead goes into the common socket.
   - The red lead goes into the socket marked with the omega symbol (Ω).

2.
   - Turn the dial to the diode testing setting.

3.
   - To calibrate the meter and ensure it's working correctly, touch the probes together. The display should show an ohms reading of close to zero.

4.
   - Like resistance testing, turn off the circuit power and disconnect the diode from the circuit.
   - Depending on the circuit design, you may need to unsolder the diode to remove it.

5.
   - Connect the probes to the diode terminals.
   - If the diode is functioning correctly, the display should show a low ohms reading in one direction only.

6.
   - When you swap the polarity of the probes, the display should show an off-limits (OL) or infinity reading.

7.
   - If it shows zero in both directions, the diode is short-circuited.

8.
   - If it shows off-limits or infinity in both directions, the diode is open-circuited.

Frequency Testing

Some multimeters come with a **frequency** testing feature.

Here's how:

1.
   - Following manufacturer's instructions, connect the test probes to the correct sockets on the multimeter.

2.
   - Turn the dial to the frequency setting, usually indicated as **Hertz (Hz)**.

3.
   - Test the component while the circuit is in operation.

Electrical Fundamentals

Temperature Measurement

Some multimeters have a temperature measurement feature. This feature usually requires an additional probe. The temperature probe typically has a separate connection socket. After setting the dial to the correct position, you can measure temperature by placing the probe end at the desired location. This feature can help you diagnose thermal management system faults, for example.

Transistor Testing

Some multimeters have a **transistor** testing feature, but automotive technicians rarely use it. Transistors are small electronic switches without moving parts. They are typically soldered to an electrical circuit board and have three connections: collector, emitter, and base or gate.

Figure 1.32 A Transistor

There are two common types of transistors: positive-negative-positive (PNP) and negative-positive-negative (NPN). Some multimeters have a transistor test feature with a six-connector socket marked PNP or NPN. To test a transistor, you need to unsolder it from its circuit and connect it to one of these sockets. Follow the multimeter manufacturer's instructions to conduct the test.

Frequency - the rate at which something occurs.

Hertz - a unit used to measure frequency. 1Hz equals one complete cycle of operation per-second.

Transistor - a non-mechanical electronic component that can function as a switch or amplifier, with no moving parts.

2-Pole Testers

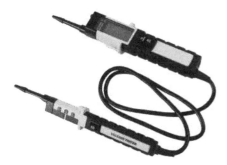

A 2-pole tester is a diagnostic tool that quickly checks if a hybrid or electric vehicle system is voltage-free. It is a simplified voltmeter with minimal settings that can automatically detect alternating or direct current without operator input. It may not be as accurate as a high-quality multimeter, but it offers a convenient way to verify that the high-voltage system has discharged below the touch threshold.

Figure 1.33 A 2-Pole Tester

Proving Unit

A proving unit is a portable device that verifies the accuracy of a voltmeter. It has a capacitor that stores a certain amount of electric charge. When the voltmeter is attached to the capacitor, it should display the same amount of charge voltage. This proves that the voltmeter is accurate. A proving unit is useful for testing high-voltage systems. It can show if the system has the right voltage by doing a 'live dead live' test.

Figure 1.34 A Proving Unit

Electrical Fundamentals

A 'live dead live' test has three steps:

- ☑ First, use the proving unit to verify the voltmeter is working (live).
- ☑ Second, use the multimeter to check the high-voltage component for absence of voltage (dead).
- ☑ Third, use the proving unit again to confirm the voltmeter is working (live).

This test ensures that the multimeter is accurate and safe.

When selecting a 2-pole tester or proving unit, make sure it is suitable for the tests you are conducting. Not all 2-pole testers have the right voltage or category rating for use with electric vehicles. They should have at least CAT III 1000 volts rating.

The proving unit should test both DC and AC currents. Also, make sure it is set to the right current type for the circuit you are evaluating.

Inductive Amps Measurement

Using an ammeter to measure electric current can be both intrusive and dangerous because it requires breaking into the circuit. Incorrect connection can also damage the ammeter. Consequently, an alternative method has been developed: some multimeters come with an inductive amp clamp, or you can purchase one separately as an add-on.

Figure 1.35 An Inductive Amps Clamp Meter

If the inductive clamp is an add-on to a standard multimeter, the wires should be connected to the voltage sockets and the voltage scale should be selected. However, the display will represent amps instead of volts (a conversion scale may need to be calculated).

Inductive amp clamps can also be used with oscilloscopes for current testing.

An amp clamp usually has a plus or minus sign indicating how it should be connected to an electrical circuit.

However, due to the shielding in many high-voltage cables and wires, inductive amp measurements may not be accurate or even possible in some electric vehicle systems.

The amp clamp uses electromagnetic interference (EMI) to measure the current flow in a circuit. It doesn't need to be connected in series but is simply clamped around the wire being tested. When the circuit is turned on and current flows, the amperage can be read from the display. (Ensure you read the manufacturer's instructions on how to connect and read the current clamp.)

While this method may not be as accurate as connecting an ammeter in series, it's faster and less likely to cause damage if connected incorrectly. Additionally, it can take much higher amperage readings than a standard multimeter.

Electrical Fundamentals

 Given the nature of high-voltage systems in hybrid and electric vehicles, it's crucial to exercise extreme caution when measuring current. If a current measurement is necessary, it's recommended to use the inductive clamp testing method, as it typically doesn't require disconnecting any high-voltage circuits.

For Low Voltage Systems:

You can test current draw at the fuse box using an inductive clamp and a small test lead with an in-line fuse. Simply replace the fuse from the circuit to be tested using the test lead (with the correct size fuse in-line). You can then clip the current clamp around the test lead and operate the circuit to get the readings.

Inductive Amp Clamp Usage:

An inductive amp clamp can quickly help determine if a non-working electrical system is due to a circuit issue or a physical mechanical fault. Attach the clamp to either the positive or negative battery lead and turn on the vehicle's ignition. Turn on the inductive amp clamp and wait for the display to stabilise. Then, operate the circuit that seems to be causing the fault.

> - If the current display changes, it suggests that a physical fault is causing the problem as the circuit is trying to do something.
> - If there's no change in the current display, it's possible that the circuit is damaged and will require further electrical testing.
> -

Megohmmeters

A Megohmmeter, crucial for diagnosing and repairing high-voltage electric vehicle systems, resembles a standard multimeter but can also conduct high-voltage resistance tests using a capacitor. It's used for **insulation** testing during high-voltage system repairs or component replacements, ensuring the high-voltage circuit is adequately **isolated** from the vehicle frame and low-voltage system.

The testing process is similar to using an ohmmeter, but instead of using the unit's low voltage battery as the source, a capacitor can be charged to different values, often up to around 1000 volts. This test aims to detect circuitry leaks, with higher voltages. Because voltage is like pressure, the higher the voltage the more likely it is to detect smaller leaks.

Figure 1.36 A Megohmmeter

 While a Megohmmeter uses minimal current that's unlikely to cause harm, the high voltages could potentially shock the operator, leading to additional injuries or accidents. Therefore, always follow recommended safety measures during an insulation test.

To prevent damage to the vehicle or its components, disconnect any sensitive electronic equipment before conducting tests, adhering to the manufacturer's guidelines.

Electrical Fundamentals

Using a Megohmmeter for Insulation Testing

Here's how:

1. - Disconnect the component from the high-voltage circuit, adhering to the manufacturer's shutdown, isolation, and dismantling procedures.

2. - Attach the probes to the Megohmmeter:
   - Insert the black lead into the common socket.
   - Insert the red lead into the insulation test socket.

3. - Calibrate the Megohmmeter as per manufacturer's guidelines for accurate measurements.

4. - Disconnect any electronic control units that could be affected or damaged by the insulation test procedure, following manufacturer's guidelines.

5. - Connect the black lead to the vehicle frame or chassis.

6. - Attach the red test lead to the high-voltage terminal and adjust the voltage scale as recommended by the manufacturer.

7. - Press the test button as per equipment instructions to apply high-voltage to the circuit.

8. - Compare your results with manufacturer's specifications..

9. - If testing a component with **Equipotential Bonding (EPB)**, move the black lead to it and repeat the test.

Equipotential Bonding (EPB) - an electrical bonding wire, often integrated with high-voltage shielding, that connects different parts of high-voltage casings. If insulation in a component is compromised, EPB aims to balance the voltage potential among other parts, reducing the risk of electrocution.

Insulation - the ability to prevent unwanted current passing other parts of an electrical system. This is essential for safety, reducing the possibility of hazards like fire or electric shock.

Isolation - the process of disconnecting a section of circuit or a piece of equipment from a source of electricity for safety reasons.

Electrical Fundamentals

Milliohm meters

A milliohm meter is a highly sensitive diagnostic tool designed to measure very low resistances in electric circuits. It's perfect for testing resistance in motor and generator phases and assessing the effectiveness of equipotential bonding (EPB).

Its usage is similar to an ohmmeter, but extra care is needed when making electrical connections to avoid inaccurate results. Also, keep in mind that a component's temperature can have a direct effect on resistance readings, so this should be considered when interpreting results.

Figure 1.37 A Milliohm Meter

Oscilloscopes

An oscilloscope is an electrical testing tool that functions similarly to a voltmeter or an ammeter. Unlike a multimeter, which may struggle to keep up with the speed of modern electronic systems in vehicles, an oscilloscope can display results in real-time.

The oscilloscope not only measures volts or amps but also time. Instead of a digital readout, it displays results as a graph of volts or amps plotted against time on a screen.

Figure 1.38 An Oscilloscope

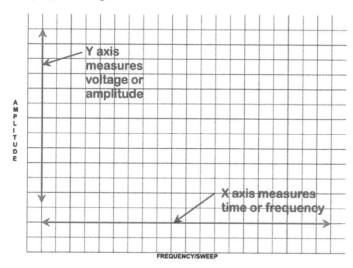

The graph typically displays voltage or amperage on the side of the screen (y-axis), often referred to as amplitude. You can adjust the number of volts or amps shown on the screen using the scale setting button or menu, depending on your oscilloscope model.

Time is usually displayed across the bottom of the screen (x-axis), often referred to as frequency or sweep. You can adjust this using the timescale button or menu, depending on your oscilloscope model.

Figure 1.39 An Oscilloscope Graph

Remembering which axis is which on a graph is easy if you think 'X is across (a cross)'.

While the numerous wires and connectors on an oscilloscope can seem intimidating, using it for basic electrical testing is straightforward. You only need two probes - a common and a voltage wire, similar to a multimeter. An inductive clamp may be needed to measure amperage.

The diagnostic sockets on oscilloscopes are often colour-coded. A quick glance at the manufacturer's instructions should tell you where to plug in these probes.

Electrical Fundamentals

Using an oscilloscope for Electrical Testing (Low-voltage circuits)

The oscilloscope probes may come in different colours, but for the sake of simplicity we will call them red and black here.

Here's how:

1. Attach the tip of the black lead to a reliable earth source, such as the battery terminal, metal bodywork, or chassis. This leaves only the red wire to manage.

2. Connect the red probe to the circuit being tested, following any manufacturer's safety guidelines. (Note: Some oscilloscopes may require an attenuator, a type of resistor, to read higher voltages).

3. Adjust the scales until an image appears on the screen.

4. With practice, you'll become familiar with the patterns and waveforms produced by different vehicle systems.

If you're unsure about the voltage or timescale settings to use on an oscilloscope, the approach is similar to using a multimeter. Start with the highest available setting and gradually decrease it until an image appears on the screen.

While an oscilloscope can be a useful tool, it is not always suitable for diagnosing and testing high-voltage electric vehicle systems. Some oscilloscopes and attachments might not meet the necessary safety requirements for working on electric vehicles. If you do conduct tests, always observe all health and safety precautions, including using high-voltage personal protective equipment (PPE).

You may also need additional probes, such as current loops or differential oscilloscope probes. Always follow the manufacturer's instructions.

Attenuators

An attenuator is a resistor component that is connected in series with the test lead and probe of an oscilloscope. It is designed to enable the oscilloscope to measure higher voltage values than those calibrated on the readout screen. Attenuators will normally be designed to reduce the displayed amplitude in multiples of ten (i.e. x10, x20 etc.) and this will be indicated on the casing of the attenuator. If an attenuator is used during your testing routine, you must multiply the voltage readings displayed on the graph by the value shown on the attenuator. However, an attenuator is not suitable to provide protection to the operator or equipment for use with electric vehicles.

Figure 1.40 An Attenuator

Electrical Fundamentals

Differential Probes

A differential probe is an oscilloscope attachment that can measure the voltage difference between two points of an electric circuit, which can often be affected by an issue known as common-mode voltage. Common-mode voltage occurs when an identical component voltage is present at both terminals of an electrical device and can interfere with the signal or cause damage to the equipment.

It consists of two input leads, a differential amplifier, and an output lead, which is connected to the oscilloscope. The integrated differential amplifier subtracts the voltages at the two input leads and amplifies the difference.

A differential probe is used to measure:

> High-voltage signals that exceed the ground-referenced input range of the oscilloscope.
> Low-voltage signals that are hidden by noise or interference.
> Differential signals used in communication or data transmission systems.
> Floating or isolated signals, similar to those generated in battery-powered devices or transformers.

Differential probes are often category rated in a similar manner to that found on a multimeter. Do not use a differential probe on an electric vehicle high-voltage system unless you have received appropriate training and are observing all relevant safety precautions and wearing the correct high-voltage personal protective equipment (PPE).

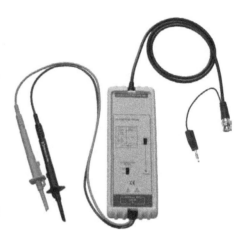

Figure 1.41 A Differential Probe

Scan Tools and Diagnostics

Diagnosing faults in modern vehicle systems often requires a scan tool, which is a device that can communicate with the vehicle's on-board computer. This is because electrical and electronic circuits operate very quickly and are difficult to test manually.

Vehicle manufacturers have been using on-board diagnostic (OBD) systems in their vehicles since the 1980s. The computers controlling the vehicle's electrical systems can self-diagnose certain malfunctions and store a code number. These electronic control units (ECUs) can record intermittent faults and store them in a keep-alive memory (KAM) for retrieval by a diagnostic trouble code (DTC) reader.

However, it's a common misconception that plugging a trouble code reader into the vehicle's OBD system will directly identify the fault. It actually only guides you towards the area of the fault, and further testing of the system and components is needed to pinpoint the actual issue.

To use a scan tool, you'll need to find the diagnostic socket. Since 2000, the type and location of this socket, also known as the data link connector (DLC), has been standardised. A 16-pin socket should be located inside the vehicle, within reach of the driver, somewhere between the centre line of the car and the driver's seat.

Data link connectors aren't always located on the driver's side of the car, often due to design considerations, especially if the vehicle was initially designed for left-hand drive.
Manufacturer's data can assist in locating the diagnostic socket.

Electrical Fundamentals

To use a scan tool, connect it to the diagnostic socket and switch on the ignition (for an electric vehicle, you may need to place it in ready mode; always follow any required safety precautions and procedures). The scan tool will then try to communicate with the vehicle's on-board computer systems. Once communication is established, follow the on-screen instructions to retrieve information and operate the on-board diagnostic system as needed.

There are two main types of diagnostic information available on many systems: OEM and E-OBD or OBD-II.

OEM refers to information from the original equipment manufacturer. To access this, you'll need to enter vehicle-specific information such as make, model, engine type, and vehicle identification number (VIN). This often provides a large amount of data specific to the manufacturer and the vehicle.

Many diagnostic trouble codes (DTCs) have been standardised:

- Codes starting with P indicate powertrain faults.
- Codes starting with C indicate chassis system faults.
- Codes starting with B indicate body system faults.
- Codes starting with U indicate network communication system faults.

However, this standard is not mandatory, so some manufacturers use their own coding system. Also, OEM information is not generic, meaning each manufacturer and vehicle type will have its own set of codes.

Scan tools typically have the following features:

- Retrieving diagnostic trouble codes from electronic control units (ECUs).
- Erasing ECU diagnostic trouble codes.
- Displaying serial data or live data.
- Showing readiness monitors.
- Resetting ECU adaptations.
- Presenting freeze frame data.
- Coding new components, if required.
- Providing information on various vehicle electronic systems.
- Resetting service reminder lights.
- Coding vehicle keys.

Many diagnostic equipment manufacturers produce scan tools that use wireless technology for communication between the data link connector (DLC) and a handheld interface used by the technician. The unit connected to the DLC, known as the vehicle communication interface (VCI), often uses a short-range Bluetooth wireless connection to transfer two-way data between a PC or dedicated scan tool and the vehicle.

Intermittent electronic faults in vehicles can be challenging and time-consuming to diagnose. Data logging, or 'flight recording', can assist in this process. This technique involves recording information from the vehicle's electronic system using scan tools that can be connected to the vehicle's diagnostic socket. The vehicle can then be operated normally, such as during a road test, while the tool captures and stores data for later review.

Several diagnostic equipment manufacturers offer specialised data logging tools. These compact devices, often no larger than a matchbox, can be connected to the data link connector (DLC) and powered directly from the serial port. They discreetly collect data while the vehicle is in use. This allows the vehicle to be returned to the customer with the data logger still attached. If a fault reoccurs, the data logger's information can be retrieved and analysed on a computer.

Electrical Fundamentals

Summary

This chapter has described:

- Basic principles of electricity and how they apply to automotive systems.
- Basic electrical tests and calculations using Ohm's law and power formula.
- Common tools and measurements for electrical diagnosis and repair.

Hybrid and Electric Vehicles

Chapter 2 Hybrid and Electric Vehicles

This chapter introduces you to hybrid and electric vehicles. You will receive an overview of different types of hybrid and electric drives, explaining their design, function, operation, and safety features. The chapter also covers other forms of alternative propulsion that differ from traditional petrol and diesel, and their impacts on the environment.

An EV or electric vehicle is an overarching term that describes any vehicle that is powered wholly or in part by an electric motor. The drive systems of these vehicles use a high-voltage energy source that poses potential dangers for those who may interact with these vehicle types.

After reading this chapter, you will have a better understanding and appreciation of hybrid and electric vehicles, and how they differ from conventional vehicles. You will also learn how to interact with these vehicles safely and efficiently, whether you are a driver, a passenger, a technician, a member of the automotive industry, or a first responder. This chapter is important for both technical and non-technical operators, as it will help you to avoid potential hazards and accidents when working on or around these vehicles.

Contents

- ❖ Context - why do we need electric vehicles **Page 39**
- ❖ Alternative propulsion **Page 44**
- ❖ Electrically propelled vehicles **Page 52**
- ❖ EV systems, components and operation **Page 60**
- ❖ Charging an electric vehicle **Page 68**

The automotive industry is a high-risk environment, especially when dealing with electrical systems. The hazards of electricity are well-known, but they can be easily ignored due to its invisible nature. This can lead to complacency unless the fundamental operation of electric vehicles is understood. Even with this understanding, caution is necessary. Do not rely on any safety systems designed for protection; instead, take precautions to minimise the risk of injury or death. Always evaluate the risks associated with any activity and implement measures to eliminate or reduce the hazards involved in any task that involves hybrid or electric vehicles.

Additional risks associated with working on, or around electric vehicles may include:

- ➤ Electrocution
- ➤ Strong magnetic fields
- ➤ Falling from height
- ➤ Injuries caused by incorrect manual handling techniques
- ➤ Chemicals
- ➤ Gases or fumes
- ➤ Hybrid engine systems starting unexpectedly
- ➤ The silent movement of electric vehicles while in use

Never attempt to work on a high-voltage electrical system unless you have received adequate training.

Hybrid and Electric Vehicles

Context - Why Do We Need Electric Vehicles

Human activities have significantly impacted our planet's environment, potentially leading to catastrophic climate changes. As our demand for natural resources grows, so does our environmental impact. Living necessitates resource use and pollution, so we must strive to operate within limits and reduce consumption where possible.

Carbon dioxide (CO_2), particularly man-made, is a significant pollutant giving rise to **climate change**. Our daily activities produce carbon dioxide, contributing to our individual **carbon footprints**. As the global population grows, so will our needs, potentially causing irreversible damage to our planet if not managed carefully.

Transportation accounts for a substantial portion of carbon emissions. Although manufacturers are innovating low-carbon technologies, individual actions play a crucial role in reducing emissions. These include maintaining our vehicles properly, avoiding unnecessary trips, walking, or cycling for short distances, using public transport, carpooling, adopting efficient driving styles, and using vehicles with alternative propulsion systems.

While no vehicle is entirely emission-free, alternative options can contribute significantly to maintaining or reducing current emission levels.

Climate Change

Global warming, an environmental phenomenon, causes the Earth's temperature to rise. While scientists agree on its occurrence, they differ on the extent, with estimates for the next century ranging from a 2°C to a 10°C increase. A rise of approximately 6°C could lead to a global extinction.

The primary cause of global warming is the **greenhouse effect**, which emanates from the electromagnetic spectrum.

The Earth receives most of its energy from the sun's solar radiation, largely in the form of visible light. This light easily passes through the Earth's atmosphere due to its wavelength and warms the ground. At night, this heat energy is radiated back towards space as infrared radiation, which has a slightly longer wavelength than visible light. Certain atmospheric gases trap this infrared radiation, contributing to global warming - a process known as the greenhouse effect.

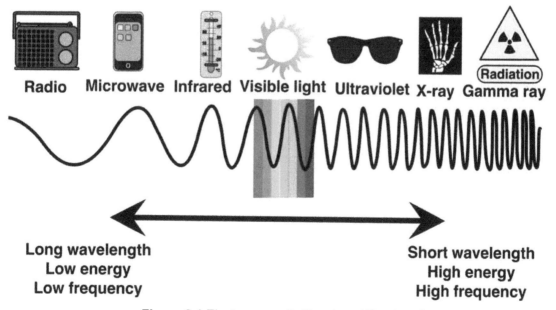

Figure 2.1 Electromagnetic Spectrum Wavelength

Hybrid and Electric Vehicles

Atmospheric gases are classified by their **Global Warming Potential (GWP)**, which represents their ability to trap infrared radiation. Carbon dioxide has a low GWP of 1, but because of its large **anthropogenic** (human-made) volume - primarily from burning fossil fuels for transportation - it poses a significant environmental issue.

The three main gases contributing to the greenhouse effect are:

- Methane
- Water vapour
- Carbon dioxide (CO_2)

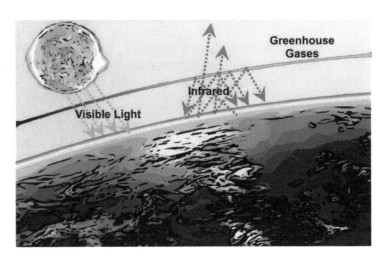

Figure 2.2 The Greenhouse Effect

Climate change - long-term shifts in temperatures and weather patterns.

Carbon footprint - the total amount of greenhouse gases, specifically carbon dioxide and methane, emitted by an activity, individual, organisation, or product.

Global warming - the long-term increase in Earth's average temperature.

Greenhouse effect - a natural process that warms the Earth's surface.

Global Warming Potential (GWP) - a measure developed to compare the impacts of different greenhouse gases on global warming.

Anthropogenic - something originating from human activity.

Termites and cows are the planet's largest methane producers.
Humans, primarily through burning fossil fuels, are the largest carbon dioxide producers, a type known as anthropogenic carbon dioxide.

Water vapour, which occurs naturally, impacts heat radiation trapping. This effect is evident in nighttime temperatures - clear nights are cooler than cloudy ones because more infrared radiation escapes into space.

Several issues have emerged in recent years because of the use of fossil fuels in vehicle propulsion systems. These include:

- The limited quantities and reserves of crude oil
- Peak oil production
- Hazardous exhaust emissions
- Environmental pollution

Hybrid and Electric Vehicles

Crude oil, the foundation for all fossil fuels used in many vehicle propulsion systems, is formed from the biological decomposition of organic materials, including plant and animal matter, under the earth's surface. This process requires extreme heat and pressure and occurs over millions of years. Crude oil provides a rich source of hydrocarbons (HC), which can be extracted and refined into fuels.

Petrol and diesel are both fuels derived from crude oil. They are produced in an oil refinery through a process called fractional distillation, in which the oil is heated, and various chemicals evaporate at different temperatures.

Figure 2.3 A Refinery

The type of fuel, whether it's petrol or diesel, is determined by the ratio of hydrogen to carbon in the fuel, which depends on the number of carbon atoms in the hydrocarbon chain. The specifics can vary slightly depending on the refining process, but generally: Petrol, also known as gasoline, typically has between five and nine carbon atoms in its hydrocarbon chain. Diesel usually has around 12 carbon atoms in its chain.

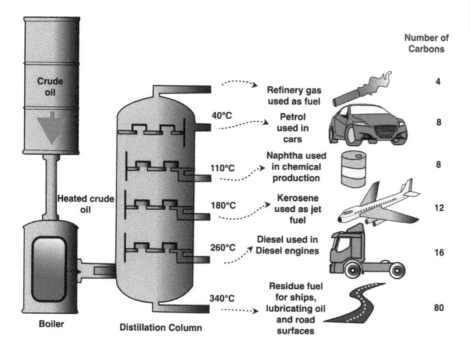

Figure 2.4 Fractional Distillation

Petrol Hydrocarbon Chain

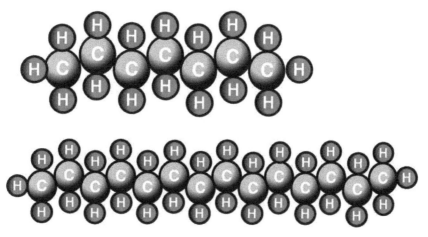

Diesel Hydrocarbon Chain

Figure 2.5 Petrol and Diesel Hydrocarbon Chains

Hybrid and Electric Vehicles

Each fuel has a **calorific value**, which is the amount of heat energy it contains, similar to how food contains calories. When the fuel burns, it releases this energy as heat that can power a vehicle. However, internal combustion engines aren't very efficient at converting this heat into other forms of energy. For instance, only about 20% of the energy from burning petrol is utilised; this means 80% of the energy is wasted.

Diesel is slightly more efficient with a potential energy usage of approximately 73%. This higher efficiency explains why a vehicle can travel further on a litre of diesel compared to the same amount of petrol.

Crude oil is a finite resource. We extract millions of barrels of oil from the Earth every day, depleting this vast energy source at a rapid pace. Once all the known reserves of crude oil are used up, it will take millions of years to replenish.

The concept of 'peak oil production' impacts the reliability of crude oil as a fuel source. As our world's power needs grow, we need to extract and process more oil to meet this demand. '**Peak oil**' refers to the maximum rate of oil production; beyond which production rates start to decline. This poses a problem as fuel shortages can occur when demand exceeds supply.

Fossil fuel - oil and fuels that naturally form over millions of years from decayed plant and animal matter.

Calorific value - the heat energy a fuel stores.

Peak oil - a situation where oil demand exceeds supply.

Exhaust Emissions and Pollution

Exhaust emissions are produced when fossil fuels, such as petrol or diesel, burn in an engine. This process is called combustion, which requires three elements: fuel, heat, and oxygen.

Fossil fuels contain hydrogen in their hydrocarbon molecules, which acts as the fuel source. The carbon in petrol and diesel, however, becomes waste after combustion. The oxygen needed for combustion comes from the air, which consists of 78% nitrogen, 21% oxygen, and 1% other gases. Since only oxygen is involved in combustion, nitrogen also becomes waste. Nitrogen is an inert gas, which means it does not burn or support combustion.

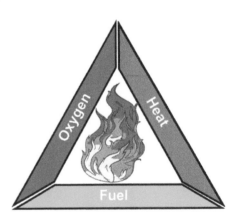

Figure 2.6 The Fire Triangle

Hydrogen, in its natural state (H_2), doesn't exist freely on Earth, except in certain underground pockets. If released, it tends to rise and escape the Earth's atmosphere.

Despite being the most abundant element in the universe, hydrogen on Earth is usually found bonded with other elements to form compounds, like the carbon in hydrocarbon fuels.

The carbon in these fuels can be likened to packaging for commercial products, which is discarded after use, leading to waste and environmental pollution.

Just as reducing or eliminating packaging is seen as an eco-friendly way to reduce waste, the idea behind alternative propulsion methods is to reduce or eliminate carbon emissions.

Hybrid and Electric Vehicles

When hydrogen and oxygen combine and are heated by a spark plug or superheated air from compression, they burn and release energy as heat. To minimise waste from combustion, the right ratio of oxygen to hydrogen is needed so that little is left after burning.

This ideal ratio is known as a stoichiometric value or a balanced chemical reaction. For petrol, this value should be 14.7:1 by mass (14.7 parts air to 1 part fuel, by weight). **Table 2.1** shows ideal air-fuel ratios for different fuel types.

Table 2.1 Ideal air-fuel ratios for different fuel types		
Fuel	**Chemical Formula**	**Air-Fuel Ratio**
Methanol	CH_2OH	6.47:1
Ethanol	C_2H_5OH	9:1
Butanol	C_4H_9OH	11.2:1
Diesel	$C_{12}H_{23}$	14.5:1
Petrol	C_8H_{18}	14.7:1
Propane	C_3H_8	15.67:1
Methane	CH_4	17.19:1
Hydrogen	H_2	34.3:1

Exhaust gases are the chemical by-products of combustion in an engine.
The primary elements involved in this process are oxygen and nitrogen from air, and hydrogen and carbon from fuel.

In an ideal scenario, combustion would only result in carbon dioxide (CO_2), nitrogen (N_2), and water (H_2O). However, combustion is rarely perfect, leading to the production of exhaust pollutants.

After fossil fuels undergo combustion, the elements recombine to form different substances, that are often considered pollutants. These substances are detailed in **Table 2.2**.

Table 2.2 Exhaust emissions and their descriptions	
Exhaust Emission	**Description**
Carbon dioxide (CO_2)	In efficient combustion, the carbon and oxygen present in the fuel and air mix to form carbon dioxide. Although not considered toxic, carbon dioxide is a greenhouse gas and is an environmental pollutant.
Hydrocarbons (HC)	If fuel passes through the combustion process without chemical change, hydrocarbons are emitted. Hydrocarbons are harmful to health and can potentially cause lung damage or cancer.
Carbon monoxide (CO)	When combustion is interrupted, due to insufficient oxygen or rapid cooling (also known as quenching), carbon monoxide (CO) is produced. This is a result of incomplete combustion. Carbon monoxide is harmful to health. It is colourless, odourless, and tasteless, but it is poisonous when inhaled. It displaces oxygen in the blood, depriving organs of the oxygen they need.

Hybrid and Electric Vehicles

Table 2.2 Exhaust emissions and their descriptions	
Oxides of nitrogen (NOx)	Combustion primarily occurs at temperatures between 2000 and 2500°C. At these high temperatures, the oxygen and nitrogen in the air combine to form a pollutant known as nitrogen oxides. Unfortunately, the more efficient the combustion process is, the more nitrogen oxides it produces. These compounds can harm lung health, damage plant life, and create photochemical smog, which reduces visibility.

To address the problems caused by fossil fuel usage, vehicle manufacturers are exploring alternative propulsion methods other than traditional petrol and diesel.

Alternative Propulsion

Liquified Petroleum Gas LPG

Liquefied petroleum gas (LPG), also known as 'Autogas', is a byproduct of the crude oil refining process. It's a low-pressure liquefied gas mixture, primarily composed of propane and butane, essentially making it a form of petrol gas. LPG can be used in standard internal combustion engines and emits less CO_2 than petrol. It is stored as a liquid under pressure, making it 250 times denser than its gaseous state.

A standard petrol vehicle can often be converted to run on LPG (a process known as retrofitting) by adding a second tank and fuel system for the LPG. The original petrol tank and fuel system remain intact, making the vehicle 'dual fuel'.

The benefits of using LPG include:

- Engines that run smoother, quieter, and cleaner than those using conventional petrol.
- An extended engine lifespan by up to 30%.
- Lower emissions output, resulting in less environmental pollution.
- Significantly lower fuel costs.
- Reduced servicing costs due to cleaner engine operation.
- Increased range if the vehicle is designed to be dual fuel (operating on either petrol or LPG).
- Potentially higher resale values as the vehicles are cheaper to run than its competitors.
- Improved crash damage safety due to the stronger construction of LPG storage containers compared to petrol tanks.
- Reduced risk of fire as LPG has a high ignition temperature, around twice that of petrol, making it less likely to spontaneously combust.

An LPG tank will never be completely empty. As the fuel is used and pressure decreases, some propane will always remain as a gas.

Additionally, an LPG tank should never be filled to its maximum capacity. It's recommended to fill the tank to about 80% of its capacity. This allows for the propane to expand and contract with changes in ambient temperatures.

Compared to traditional petrol or diesel vehicles, those running on liquefied petroleum gas (LPG) produce emissions that are significantly less harmful to both health and the environment.

Hybrid and Electric Vehicles

Table 2.3 provides a comparison of how petrol and diesel stack up against LPG in terms of emissions.

Table 2.3 Emissions comparison between petrol, diesel and LPG	
LPG compared to petrol	**LPG compared to diesel**
75% less carbon monoxide	60% less carbon monoxide
40% less oxides of nitrogen	90% less oxides of nitrogen
87% less potential of forming ground level ozone	70% less potential of forming ground level ozone
85% less hydrocarbons	90% less particulate matter (soot)
10% less carbon dioxide	

Compressed Natural Gas CNG

Compressed natural gas (CNG) is an alternative to LPG. It's primarily composed of methane and can be produced from various sources, including the 3% of natural methane gas found in crude oil. CNG can fuel standard internal combustion engines as a substitute for petrol. Methane combustion produces the least CO_2 of all fossil fuels. Like LPG, petrol cars can be retrofitted to run on CNG, becoming bi-fuel natural gas vehicles (NGVs).

The CNG system operates similarly to the LPG system, but the fuel is stored as a gas instead of a liquid. Storing **methane** as a liquid would require not only high pressures but also extremely low temperatures. A drawback of CNG compared to LPG is that due to the lower density of the stored fuel, CNG vehicles need larger fuel tanks to store similar quantities. However, an advantage is that compressed natural gas can be stored at much lower pressure in a form known as adsorbed natural gas (ANG), and vehicles can be refuelled from the regular natural gas network without further compression.

Biogas typically refers to a gas produced by the biological breakdown of **organic matter** in the absence oxygen. It usually comes from **biogenic** material like dead plant and animal matter and is a type of renewable biofuel. A **biogas** generator plant typically contains an **anaerobic** digester where plant material, animal matter, and sewage decompose to form methane, carbon dioxide, and small amounts of hydrogen sulphide. After purifying the raw gas, compressed biogas can power standard internal combustion engines similarly to CNG.

Figure 2.7 An Anaerobic Digester

Methane - a colourless, odourless flammable gas that is the simplest member of the paraffin series of hydrocarbons.

Organic matter - material that originates from living organisms and can decay.

Biogenic - produced by living organisms.

Biogas - a renewable energy produced from the fermentation of organic matter.

Anaerobic - occurring without oxygen.

Hybrid and Electric Vehicles

Biodiesel

Biodiesel, also known as fatty acid methyl ester, is a type of diesel fuel derived from a variety of oil-producing plants. This provides a more sustainable alternative to traditional diesel fuels, as it reduces greenhouse gas emissions, improves engine performance, and supports local agriculture.

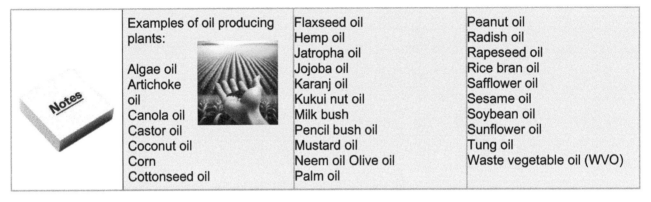

Examples of oil producing plants:	Flaxseed oil	Peanut oil
	Hemp oil	Radish oil
	Jatropha oil	Rapeseed oil
Algae oil	Jojoba oil	Rice bran oil
Artichoke oil	Karanj oil	Safflower oil
	Kukui nut oil	Sesame oil
Canola oil	Milk bush	Soybean oil
Castor oil	Pencil bush oil	Sunflower oil
Coconut oil	Mustard oil	Tung oil
Corn	Neem oil Olive oil	Waste vegetable oil (WVO)
Cottonseed oil	Palm oil	

Table 2.4 provides a comparison of the advantages and disadvantages of biodiesel versus traditional diesel fuel. This comparison can help illustrate the potential benefits and drawbacks of using biodiesel as an alternative to fossil diesel.

Table 2.4 The advantages and disadvantages of biodiesel	
Advantages of biodiesel	**Disadvantages of biodiesel**
Biodiesel, primarily derived from plants or crops, is considered a renewable energy source, unlike fossil fuels.	Biodiesel has a lower energy density compared to traditional diesel fuel. As a result, vehicles running on biodiesel may not achieve the same fuel efficiency as those powered by conventional diesel.
Biodiesel is often seen as **carbon neutral** when used as fuel for vehicles. This is because the carbon dioxide emitted during its burning is balanced by the amount absorbed from the air during the plant's growth and lifespan.	Making biodiesel can be costly and energy intensive. The energy required often comes from non-renewable sources, which results in carbon dioxide emissions during production.
Biodiesel is fully **biodegradable** and harmless, so fuel spills pose less environmental pollution risk than fossil fuels.	Some biodiesels, particularly those made from waste vegetable oil, contain chemicals that can harm natural rubber, potentially causing leaks in fuel system seals.
Biodiesel's **flash point** is higher than that of fossil diesel, which lowers the chance of fire in case of an accident.	Large oil-producing crops require vast land areas. This often results in less space for food crops or leads to deforestation.

Flash point - the temperature at which a fuel can ignite in air.

Carbon neutral - a state where there is a net zero release of carbon dioxide into the atmosphere.

Biodegradable - something that is capable of being decomposed by bacteria or other living organisms, avoiding pollution.

Hybrid and Electric Vehicles

Bio-alcohol/Ethanol

Various types of alcohol can be used as fuel for internal combustion engines, but mainly methanol and ethanol. Methanol, often made from natural gas, isn't usually considered a biofuel, though it can be produced from biogas, which is more complex and costly.

Ethanol is primarily produced from biological material through fermentation, similar to the process used in making alcoholic beverages.

Ethanol can also be made from petroleum-based products, but it's unsafe to drink and can cause blindness or death.

Alcohols used for fuel have a lower energy density than conventional petrol or diesel, requiring about 1.5 to 2 times the volume to produce the same amount of heat energy. However, alcohol's high **octane** allows it to operate at higher compression pressures in the engine, resulting in similar performance and fuel economy. It also reduces tailpipe emissions compared to petrol or diesel.

A downside of alcohol is its tendency to corrode or cause corrosion in fuel system components over time. This could lead to excessive corrosion or blockages in fuel systems. Many car manufacturers have been designing cars with fuel systems that can tolerate about 10% ethanol, often blended with standard petrol. But to run solely on alcohol, fuel system materials need to be adapted and engine management systems reprogrammed.

Diesel engines can also run on alcohol, but because alcohol has a low **cetane** value, additives like glycol are needed to improve ignition.

Octane - a colourless, combustible hydrocarbon that helps prevent detonation.

Cetane - a measure of a fuel's ignition delay, or the time from injection until it begins to burn.

Other alcohols like butanol and propanol can also be used as fuel. However, they are more costly and harder to produce than methanol and ethanol, making them less suitable for fuel production. If an economical method to produce butanol is found, it could improve fuel economy and performance, as its energy density is very similar to petrol.

Ammonia Green

Ammonia Green (NH_3) is successfully used by some vehicle manufacturers as it can operate in spark ignition or compression ignition engines with only minor changes required.

Using ammonia as a fuel for internal combustion engines offers several benefits:

- It has a high energy density compared to other non-petroleum-based fuels.
- It doesn't contain carbon, so it doesn't release carbon dioxide or generate particulate matter during combustion.
- Since ammonia can be easily made, it won't deplete like fossil fuels.

Hybrid and Electric Vehicles

Ammonia is produced by reacting hydrogen with nitrogen from the air to create NH_3, meaning the raw materials needed are water and air.

The most common way to create hydrogen for ammonia production is from natural gas, a process that can release significant greenhouse gases. This is called brown ammonia.

If hydrogen for ammonia production is created through electrolysis, non-polluting methods can be used, including:

- Wind power
- Solar
- Hydroelectric
- Wave power

Figure 2.8 Renewable Electricity Sources

If ammonia is produced through electrolysis from renewable energy sources, it's known as 'ammonia green'. Liquid ammonia has half the density of petrol or diesel, making it easy to carry in sufficient quantities in vehicles and provide a good driving range. It produces no emissions other than nitrogen and water vapour upon combustion, making the exhaust non-polluting.

 While ammonia is considered highly toxic, it's no more dangerous than petrol or LPG when stored and handled properly.

Synthetic Fuels

Synthetic fuel is a man-made alternative to fossil fuels. It's a type of hydrocarbon that can be converted into petrol or diesel. Here's how it's made:

1. Water Separation: Water is split into hydrogen and oxygen through a process called **electrolysis**.
2. **Syngas** Creation: The hydrogen is then mixed with carbon dioxide (CO_2) to create a gas mixture known as syngas.
3. Fuel Conversion: The syngas is processed into a liquid form, producing synthetic petrol or diesel.

One of the benefits of synthetic fuel is that it uses CO_2, which can be sourced from other carbon-emitting technologies. Although this carbon is released back into the atmosphere when the fuel is burned, it's still considered more eco-friendly than fossil fuels, as it reduces the net carbon footprint of the fuel cycle.

However, there's a downside: the process requires a significant amount of electricity. This is used to split water into hydrogen and oxygen, and to combine these with CO_2 to create syngas. Unfortunately, this process is quite inefficient and results in a lot of energy being wasted.

Hybrid and Electric Vehicles

Hydrogen

Hydrogen is a clean and eco-friendly energy carrier. However, it's important to note that hydrogen in its natural state (H_2) would escape Earth's atmosphere if not contained. Here's how we use it:

1. Separation: Hydrogen is often separated from other compounds through electrolysis or chemical reactions.
2. Storage: Once separated, it can be stored in containers and used as a portable energy source.

Since energy is needed to extract hydrogen, it's not considered a fuel but rather an energy carrier, similar to a battery. The key difference is that when the energy is converted back (for example, to power a vehicle), it combines with oxygen from the air and the only byproduct is water. This means no harmful carbon is released into the atmosphere when in use. However, the production of hydrogen and its source compound may release carbon during processing.

While hydrogen is naturally colourless, colour codes are often used to describe its source material and production method. These colour codes are described in **Table 2.5**.

Table 2.5 Hydrogen colour codes	
Colour Code	**Description**
H_2 Black or Brown	The extraction of hydrogen from black coal or lignite, also known as brown coal, is considered to be highly polluting. The process, known as gasification, involves several steps. Initially, the coal is dried and then subjected to **pyrolysis**, a heating process that breaks it down into different components. Some of this material is then burned to further 'crack' or break it down. Finally, steam is introduced to produce syngas and carbon dioxide (CO_2).
H_2 Grey	Grey hydrogen, which is highly polluting, is produced from natural gas, primarily composed of methane (CH_4). This process involves combining methane with steam at extremely high temperatures in a procedure known as **'steam reformation'**. This reaction results in the production of hydrogen (H_2) and carbon dioxide (CO_2). The hydrogen is stored for future use, while the CO_2 is released into the atmosphere, contributing to the greenhouse effect.
H_2 Blue	Blue hydrogen is produced in the same way as grey hydrogen, where methane (CH_4) is converted into hydrogen (H_2) and carbon dioxide (CO_2) through a process called 'steam reformation'. The key difference is that the CO_2 produced in blue hydrogen production is captured and stored instead of being released into the atmosphere as a **greenhouse gas**.
H_2 Green	Green hydrogen is produced through a process called electrolysis, which uses clean, renewable energy to extract hydrogen from a source material. This renewable energy can come from wind, solar, or hydroelectric power, often when there is an excess of energy being generated. If the energy used is exclusively from solar power, the hydrogen produced is sometimes referred to as 'yellow'. As this method utilises renewable energy sources, it is considered the most environmentally friendly way to produce hydrogen.
H_2 Pink	Pink hydrogen is produced using electrolysis, similar to green hydrogen. However, the key difference is that the electricity used in this process is generated from nuclear power. Nuclear power does not emit carbon dioxide into the atmosphere during electricity production, making it a cleaner option.

Hybrid and Electric Vehicles

Hydrogen extraction requires energy, so it's more accurate to consider hydrogen as an energy carrier rather than a fuel.

Storing and transporting hydrogen presents challenges. It requires high pressures or extremely low temperatures to store adequate quantities because hydrogen, in its gaseous state, occupies a large volume. Specialised containers are necessary to prevent natural leakage over extended periods.

Synthetic - something that is artificially made, often copying a natural product.

Electrolysis - a process where electric current is passed through a substance creating a chemical change.

Syngas - short for synthesis gas, is a mixture of hydrogen and carbon dioxide.

Pyrolysis - the thermal breakdown of substances at high temperatures.

Greenhouse gas - a type of gas that absorbs or emits infrared radiation, trapping heat and contributing to climate change.

Hydrogen Internal Combustion Engines (HICE)

Hydrogen gas can be used to power internal combustion engines with minor modifications. The benefits of using hydrogen include the emission of only water vapour upon combustion and the potential for higher power output when compared to petrol or diesel.

While it's not challenging to create a hydrogen-powered engine, optimising its performance is more complex.

Modifying the fuel delivery system is typically necessary for an engine to operate on hydrogen. There are three primary methods for introducing hydrogen:

- Single-point injection.
- Multi-point injection.
- Direct injection.

Single-point and multi-point methods require minimal engine adaptation, and hydrogen easily converts into a gas within the intake manifold. Direct injection might necessitate significant engine modifications but provides the best operational range when running on hydrogen, as it allows for more detailed adjustments of the air-fuel mixture.

The ignition system of a hydrogen internal combustion engine can resemble a standard petrol ignition system, although cold-running spark plugs are needed to prevent **pre-ignition**. However, platinum electrode spark plugs should not be used as this metal acts as a **catalyst** and reacts with hydrogen, causing it to oxidise with air. This can lead to engine damage and reduced performance.

Pre-ignition - a situation where the fuel in the combustion chamber ignites before the spark is generated.

Catalyst - A component that starts and maintains a chemical reaction but is unaffected by the process.

Hybrid and Electric Vehicles

While hydrogen is a beneficial energy source for vehicles, its characteristics present both pros and cons when used in internal combustion engines.

Table 2.6 describes some advantages and disadvantages of using hydrogen in internal combustion engines.

Table 2.6 The advantages and disadvantages of hydrogen when used in HICE		
Property	**Advantages**	**Disadvantages**
Its wide range of flammability.	Hydrogen can combust even with very lean mixtures that are significantly below the recommended air-fuel ratio. This property facilitates easy engine starting and promotes more thorough combustion.	There's a limit to how lean an air-fuel mixture can be before the combustion temperature drops to a level where it diminishes the engine's power output.
Its low ignition energy requirements.	Hydrogen requires significantly less energy to ignite compared to petrol. This allows a standard spark ignition system to ignite the fuel, even with extremely lean mixtures.	Due to hydrogen's low ignition energy requirements, hot spots within the cylinder can trigger pre-ignition, resulting in a misfire. These hot spots could be caused by carbon build-up from burnt lubrication oil or overheated spark plug electrodes.
Its ability to resist quenching.	Hydrogen, unlike petrol, is less likely to extinguish when it comes into contact with the cooler cylinder walls during combustion. This means it can use more of its available energy.	The small distance at which hydrogen stops burning can increase the chance of a backfire. This is because the burning hydrogen can get closer to an almost closed inlet valve. If the fuel system uses single-point or multipoint injection in the manifold, there's a risk that fuel could ignite and cause a fire.
Its high auto ignition temperature.	A high auto ignition temperature contributes to the fuel's stability, preventing it from detonating just because of the air temperature. This stability allows it to operate at much higher compression ratios than a typical petrol engine. As compression ratios increase, so does the engine's performance.	Hydrogen's high auto ignition temperature makes it unsuitable for use in engines that ignite by compression.
Its ability to spread out due to its low density.	Since hydrogen remains a gas even at very low temperatures, it will quickly spread and disperse if there's a leak in the system. This reduces the chance of an accidental fire.	Hydrogen's low density means it occupies more space in an engine cylinder than vapourised petrol. As a result, there's less energy density available for combustion, which can decrease the engine's performance.

The ideal air-fuel ratios for a petrol engine and a hydrogen engine are 14.7:1 and 34:1 by mass respectively. However, hydrogen engines often operate at even lower ratios, down to as low as 180:1

Hybrid and Electric Vehicles

Hydrogen is highly flammable. If an accidental leak in a hydrogen system should occur, ensure that all sources of ignition are removed.

Vehicles are often categorised based on the emissions they produce. An Ultra-Low Emission Vehicle (ULEV) is one that emits a minimal amount of carbon dioxide (CO_2) while in operation. On the other hand, a Zero Emission Vehicle (ZEV) doesn't produce any CO_2 emissions while it's being driven. Even though some vehicles might not have an exhaust system, the term 'emissions produced at tailpipe' is often used to refer to these categories.

Electrically Propelled Vehicles

What Makes an Electrically Propelled Vehicle

An electric vehicle is one that uses electricity, typically from an onboard battery or a hydrogen fuel cell, for all or part of its propulsion. Electric cars have been around for a long time. They're often credited to various inventors, but the main credit goes to a Hungarian inventor named Ányos Jedlik. In 1828, he created a small model car equipped with an electric motor he designed. By 1838, a Scotsman named Robert Davidson had built an electric locomotive that could reach a top speed of 4 mph. These designs were around nearly 50 years before the first cars with internal combustion engines.

When working with hybrid and electric vehicles, it's crucial to be aware of the dangers of electricity, including the risk of electric shock or electrocution. Since electricity is invisible, it can only be detected with specialised equipment. Here are some key electrical units to understand:

- **Voltage**: This is the electrical pressure or force.
- **Amperage**: Also known as electric current, this is the quantity of flowing electricity.
- **Resistance**: This is a restriction in electric current flow, measured in ohms.
- **Power**: This is the rate at which work is done, measured in watts.

All these units influence the risk of electric shock or electrocution. However, voltage is the primary factor to consider.

The human body naturally resists electrical voltage up to a certain limit. When this limit is exceeded, the voltage induces a potentially harmful current flow. It's important to note that while voltage can be dangerous, it's the current that causes damage.

The level of voltage considered dangerous varies among individuals due to differences in personal electrical resistance and points of contact with an electrical circuit. Therefore, various values may be quoted throughout the book, but caution is advised when dealing with any voltage.

For dry, unbroken human skin, the touch threshold is often cited as 50 volts DC or 25 volts AC. This value can decrease if the skin is wet or if the points of electrical contact penetrate the skin. Once the voltage touch threshold is reached, a current as small as 80 milliamps (a milliamp is one-thousandth of an amp) can be fatal under certain circumstances.

Hybrid and Electric Vehicles

According to vehicle regulations, ECE R-100 paragraph 2.14 defines high-voltage as a classification for an electrical component or circuit. It's considered high-voltage if its working voltage exceeds 60 volts DC but is less than 1500 volts DC, or if it's over 30 volts AC but less than 1000 volts AC RMS.

This implies that the voltage found in the drive and propulsion systems of electric and hybrid vehicles can be extremely dangerous. The average voltage used for these drive systems typically ranges between 100 and 650 volts DC. However, some manufacturers may use significantly higher voltages in their design and construction of electric vehicles, with several using at least 800 volts.

Short circuit

Electrocution can occur at the stated voltage level due to the internal resistance of the human body. However, a short circuit, which can be created by accidentally bridging the high and low voltage potential with a metal conductor, can cause an immediate discharge of electric current. This can result in extreme heat, sparks, fire, and even explosions. This can happen even if the voltage potential has dropped below the safe touch threshold. Therefore, electrical voltage is potentially dangerous. When working around these systems, it's recommended to use insulated tools to reduce the risk of accidental short circuits.

Magnetic fields

Electric vehicles feature strong permanent magnets and generate magnetic fields around high-voltage cables when in use. These intense magnetic fields can pose a risk to individuals who rely on electronic life-sustaining devices such as pacemakers, automatic defibrillators, and insulin pumps. Although these devices are well-shielded, making it generally safe for individuals to operate and drive these vehicles, dismantling the vehicle for access to different components could potentially endanger people with certain health conditions. Therefore, it's advised that individuals with electronic life-sustaining devices avoid working on the high-voltage propulsion systems of electric and hybrid vehicles.

Hybrid and Electric Vehicles

Electric vehicles fall into two main categories. First, there are hybrid vehicles that use both an internal combustion engine and an electric motor for propulsion. Second, there are pure electric vehicles that rely solely on an electric motor, powered by either an onboard battery or a hydrogen fuel cell.

These two categories can then be broken down into further subsections:

Hybrid

Hybrid vehicles have two power sources - typically an internal combustion engine and an electric motor. Some are designed primarily as traditional vehicles with an engine, supplemented by an electric motor for enhanced performance, lower emissions, and better fuel economy. Others are designed primarily as electric vehicles with the main propulsion from an electric motor, and an engine used as an onboard generator to supplement the electricity supply.

Plug-in Hybrid Electric Vehicles PHEV

A plug-in hybrid is a vehicle with a larger onboard battery that can be charged from an electrical outlet. This allows it to have a limited all-electric driving range. Once the battery's energy is depleted, the vehicle operates like a regular hybrid. This category also includes two design types: one is primarily an engine-based vehicle supplemented by an electric motor, and the other is primarily an electric vehicle with an onboard generator.

Hybrid and Electric Vehicles

Extended Range Electric Vehicles E-REV

An extended-range electric vehicle, also known as a series hybrid, is primarily driven by an electric motor. It also has an onboard internal combustion engine that acts as a generator, providing additional range when needed.

Electric Vehicles EV

The term 'electric vehicles' broadly refers to any vehicle that uses an electric drive for propulsion. This category encompasses a wide range of vehicles with electric drives, including but not limited to scooters, bicycles, motorcycles, mopeds, cars, vans, lorries, buses, trucks, forklifts, golf buggies, and so on.

Pure Electric Vehicles PEV

A pure electric vehicle is one that relies solely on an onboard battery, charged from an electrical outlet, for propulsion. A subset of this category includes hydrogen fuel cell vehicles. These are also purely electric, but the electrical energy is generated by combining hydrogen and oxygen in a fuel cell to produce water, with the excess energy being converted into electricity.

Hydrogen Fuel Cell HFC

A Hydrogen Fuel Cell (HFC) vehicle is a type of electric vehicle. It operates similarly to a pure electric vehicle but uses a different energy source. Instead of relying on a large, heavy, high-voltage battery for power, it uses hydrogen and oxygen to produce electricity.
In an HFC vehicle, a device called a fuel cell combines hydrogen from an onboard tank with oxygen from the air. This process, facilitated by a catalyst, bonds the hydrogen and oxygen atoms to form water. The reaction produces heat and an electric current as byproducts.
The generated electricity can be used in two ways: it can either charge a small high-voltage battery that powers the vehicle's propulsion system, or it can be directed straight to the drive motor.
One of the main advantages of HFC vehicles is their refuelling process. Unlike traditional electric vehicles that require charging from an electrical outlet, HFC vehicles are refuelled much like petrol or diesel vehicles. When the hydrogen supply runs low, the driver simply refuels at a pump, allowing for continued operation.

Electric Quadricycles

The electric quadricycles category includes a variety of electric vehicles that don't fit into other categories. These often include amateur-built vehicles or models that are small, with restricted power and speed. They are not subject to the same rigorous testing as standard road cars.

A solar vehicle is powered by energy from the sun, which is obtained from solar panels, also known as photovoltaic cells. This term comes from the words 'photo' meaning light and 'voltaic' meaning electricity. Photovoltaic cells are made of semiconductor material such as silicon. When sunlight strikes the semiconductor material, some of the energy is absorbed and is converted into moving electrons within the material to create an electric current. The flow of electrons is in one direction (direct current). If electrical contacts are placed above and below the photovoltaic cell, the electricity produced by converting photons can be used to power electrical circuits. The silicon used in a photovoltaic cell works in this way because its atoms are not completely filled up with electrons in their outer orbit or shell. If impurities are added to the material, such as phosphorus, small amounts of extra electrons are available, making it negative. On the other hand, if boron is added to the silicon, fewer electrons are available, making it positive. These two sections of silicon can then be joined, and if supplied with light from the sun, share the electrons between the positive and negative sections, creating electric current.

Hybrid and Electric Vehicles

A solar panel converts light energy into electricity that can be used as a source of power. The sun provides approximately 1000 Watts of energy for every square metre of the earth's surface. Unfortunately, not all of the energy provided by the sun can effectively be converted into electricity. This is because light comes in different wavelengths and photovoltaic cells can only use certain areas of the electromagnetic spectrum. Apart from some specialist designs, solar panels are not used to directly supply a vehicle with enough power to provide drive to electric motors, but they can be used to charge batteries or extend the range of 'plug in' electric vehicles. To enable the use of solar power to provide electric drive, most of the exposed surface of the vehicle will be needed for photovoltaic cells. Only the cells directly in sunlight will get full energy, and the buildup of dirt will also affect efficiency. The added cost of components and electronics makes this design unsuitable in most situations, and potential accident damage is a factor that must be considered when positioning solar panels.

These categories and subcategories are designed to distinguish different types of electric vehicles. However, due to manufacturers' specialisations and unique designs, it can be challenging to categorise all makes, models, and types. Early electric vehicles were often easy to identify due to their unique designs for aerodynamics and economy. With technological advancements, most hybrid and electric vehicles now externally resemble traditional petrol and diesel models. However, there are several identifying features that can indicate the use of an electric drive.

Identifiers

Usually, the first identifier is badges with the words EV, electric vehicle, or hybrid visible on the vehicle's exterior. However, this is not always the case.

Figure 2.9 Hybrid and EV Badges

Hybrid and Electric Vehicles

To differentiate between hybrids, fully electric vehicles, and plug-in hybrids, look for electric charging ports (sometimes hidden behind badges or flaps) on pure electric vehicles. Hybrid electric vehicles may have fuel filler caps and exhaust pipes (which are not required on a pure electric vehicle), or a combination of both on plug-in hybrid electric vehicles.

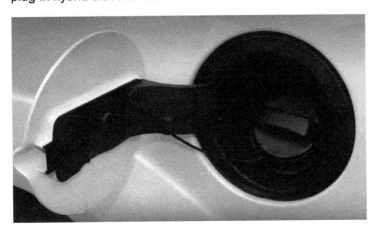

Figure 2.10 Fuel Fillers and Exhaust Pipes

The driver's display inside the vehicle often provides clues about whether the vehicle is electric or hybrid. It usually shows energy distribution or other indications of electric operation, consumption, and range. Note that there should be some indication that the car is in ready mode to inform the driver that the car is live and could move unexpectedly.

Figure 2.11 Drivers Display

Since electric drive does not require a traditional gearbox, most hybrids and electric vehicles typically have some form of automatic transmission. Therefore, the drive selector may resemble that of a traditional automatic vehicle.

Figure 2.12 EV Gear Selector

Hybrid and Electric Vehicles

In what would be considered the engine compartment of a traditional vehicle, there may also be badges stating hybrid or electric drive, high-voltage warning labels indicating dangers, and brightly coloured (often orange) high-voltage wiring.

Figure 2.13 EV Engine Bays

Within the class of hybrid vehicles, there are several subcategories.

Micro Hybrids

A micro hybrid typically has a motor generator power of between one and two kilowatts. The level of hybridisation is usually less than 4% and often involves a belt-driven or crankshaft support from the electric motor generator. The assistance is generally limited to a Start-Stop function, with minimal regenerative braking capability. The working voltage for micro hybrids is typically around 14 volts DC.

Mild Hybrids

A mild hybrid typically has a motor generator power of between 5 and 20 kilowatts. The level of hybridisation is approximately 3 to 20%, often using a crankshaft-mounted electric machine. This machine provides torque assist and regenerative braking but does not support an electric-only mode. Mild hybrids operate at a medium to high-voltage, ranging from 48 to 100 volts DC.

Full Hybrids

A full hybrid typically has a motor generator power of between 20 and 60 kilowatts. The level of hybridisation is approximately 20 to 40%, offering a significant boost to torque and regenerative braking. Full hybrids often have an electric-only mode, but the range is limited due to a relatively small battery. The typical working voltage for a full hybrid is usually above 200 volts DC.

Plug-in Hybrids

A plug-in hybrid typically has a motor generator power of between 40 and 80 kilowatts. The level of hybridisation ranges from 40 to 100%, offering high boost and torque with regenerative braking. Plug-in hybrids have an electric-only mode, with a longer range than a full hybrid due to a slightly larger battery and the ability to charge from an electrical outlet. The typical working voltage for plug-in hybrids is usually above 200 volts DC.

Hybrid and Electric Vehicles

Drivetrains

The drivetrain or powertrain of hybrid electric vehicles typically follow one of four designs. A simplified, non-technical overview of these designs is provided in the following section.

Parallel

A parallel hybrid uses a small internal combustion engine, which offers benefits such as low emissions and good fuel economy. The engine's performance is enhanced by an electric motor connected to the crankshaft, which boosts power output using electricity. The engine and electric motor always operate together side-by-side, hence the term 'parallel'.

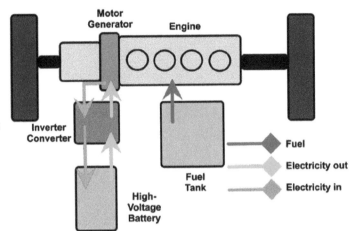

Figure 2.14 Parallel Hybrid Layout

Series

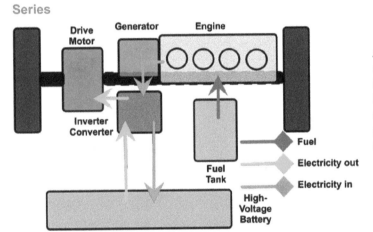

Figure 2.15 Series Hybrid Layout

A series hybrid is primarily designed as an electric vehicle, but its battery capacity is smaller than that of a full pure EV. To extend the battery's range, an internal combustion engine is included to act as an onboard generator, supplementing the electricity supply as needed.

Power Split

A power-split hybrid merges the benefits of both series and parallel designs. It includes an internal combustion engine and two motor generators. One motor generator is connected to the engine crankshaft, while the other is separate and can drive the wheels like an electric car. A specialised planetary gear set is positioned between the two motors, providing the ability for the motor generators to operate independently at different speeds. Depending on which electric motor is powered, the vehicle can operate in series, parallel, or a combination of both modes. The driver doesn't need to take any action, as the vehicle's onboard computer automatically selects the most efficient mode.

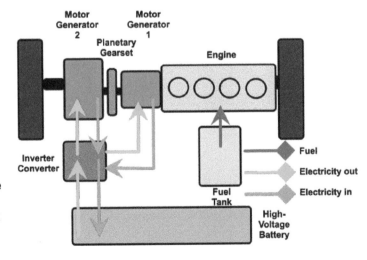

Figure 2.16 Power Split Layout

Hybrid and Electric Vehicles

Dual

A dual hybrid utilises a four-wheel drive structure. The internal combustion engine drives one axle, while an electric motor drives the other. This setup allows the vehicle to operate in engine-only mode, electric drive mode, or in four-wheel drive mode using a combination of both. Charging of the high-voltage drive battery is achieved through **regenerative braking** and recovery only.

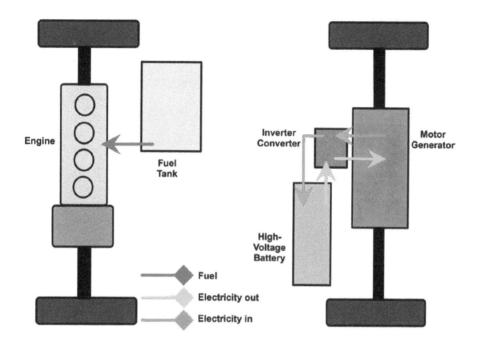

Figure 2.17 Dual Layout

Vehicle designers are continually innovating unique hybrid drive systems. These innovations not only enhance efficiency and performance, but also avoid **copyright** issues.

Manufacturers sometimes use the terms **'series'** and **'parallel'** differently when explaining their hybrid drive systems. It's crucial to thoroughly understand the manufacturer's explanation of their system before making assumptions about its operation. This approach helps avoid confusion and misunderstandings.

Regenerative braking - a way of capturing some of the kinetic energy that is normally lost when a vehicle slows down or stops.

Copyright - refers to the exclusive legal right to own and use a design.

Series - describes items connected in a sequence, one after another.

Parallel - refers to items connected side by side.

Hybrid and Electric Vehicles

EV Systems Components and Operation

Figure 2.18 Ready Mode

Hybrid and micro hybrid vehicles often come with auto stop systems to enhance fuel efficiency and reduce emissions. These systems automatically stop the engine when the vehicle is stationary, eliminating unnecessary idling. The engine restarts automatically based on the vehicle's needs and the driver's actions. Factors influencing the restart include brake pedal pressure, engine temperature, battery charge, cabin temperature, and air conditioning needs.

The dashboard typically displays an indicator when the vehicle is in 'ready mode', signifying that the engine may restart at any moment. This is crucial for safety reasons, especially when working on or around the engine and electrical system. The high-voltage system could still be live, and the engine may start unexpectedly.

For fully electric vehicles, 'ready mode' means the vehicle is powered up and could operate without warning. This could pose a risk as the vehicle might move unexpectedly. Therefore, when working on or around hybrid or electric vehicles, it's important to ensure 'ready mode' is turned off. This can be done by switching off the ignition and keeping any smart key out of its operational range.

Signs and barriers should be implemented to alert others that work is being conducted on an electric vehicle and to indicate any potential hazards.

To prevent unauthorised interference with the vehicle's systems, specialised lockout procedures could be employed. This ensures that the vehicle's systems remain secure during maintenance or repair work.

Hybrid and Electric Vehicle Regenerative Braking

A standard braking system uses friction to transform a vehicle's kinetic (movement) energy into heat. However, hybrid and electric vehicles often employ a method called regenerative braking to assist in braking and deceleration. This efficient process converts some of the vehicle's kinetic energy into electricity, which can then be used to charge the high-voltage battery system.

Hybrid vehicles use a combination of an internal combustion engine and an electric motor for propulsion. In contrast, fully electric vehicles rely solely on a drive motor. If the electric motor is mechanically driven, such as during braking, it can function as an electrical generator. This conversion of kinetic energy into electricity helps slow down the vehicle. Any additional deceleration needed by the driver that is not provided by regenerative braking is managed by traditional friction brakes, using brake callipers and pads against discs. A sophisticated electronic control system calculates the required amount of braking and divides the operation between the generator and the brakes.

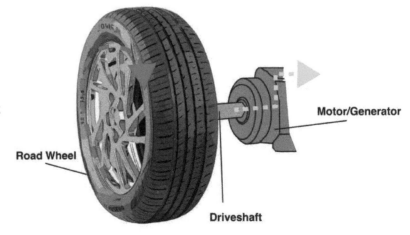

Figure 2.19 Regenerative Braking

Hybrid and Electric Vehicles

Regenerative braking has a few limitations:

1. The regenerative braking effect decreases at lower speeds.
2. Most road vehicles with regenerative braking only power some wheels, such as in a two-wheel drive car. Regenerative braking power is only applied to the drive wheels because they are the only ones connected to the drive motor. To ensure controlled braking under challenging conditions like wet roads, friction-based braking is needed on the other wheels.
3. If the batteries or capacitors are fully charged, regenerative braking does not occur.

Hybrid and electric cars with regenerative braking experience less wear on their brakes compared to conventional vehicles. This is because the electric motor helps slow the vehicle down, reducing the need for traditional brakes. However, if drivers rely too much on regenerative braking and don't use the standard brakes enough, it can lead to brake failure or seizure of braking components.

Operation of Hybrids Under Various Modes

Hybrid vehicles can function differently based on driving conditions due to their design. Here are some common drive modes:

- **Start** - The electric motor, which provides maximum torque, is often used alone or with the engine for strong acceleration and less fuel use when starting off.
- **Low Speed** - The engine stops and only the electric motor drives the vehicle at low speeds.
- **High Speed** - During acceleration or high-speed driving, the vehicle may run solely on engine power and the electric motor is off.
- **Overtake** - For rapid acceleration, both the engine and electric motor power the vehicle for maximum performance.
- **Slowing** - When slowing down or braking, the engine stops, and the electric motor recharges the batteries by acting as a generator.
- **Stop** - The engine and electric motor automatically turn off to save power when the vehicle is stationary.

Electric Components That Make up the Hybrid and Electric Vehicle System

The following is a description of common components which make up the electrical propulsion systems of both hybrid and electric vehicles.

Ignition Switch

In most hybrid and electric vehicles, a wireless **smart key** is used for the access and starting procedures. The key doesn't need to be inserted into a slot and turned. Instead, a push-button start is often used. This button works similarly to traditional ignition keys and has three positions:

1. First Press - Activates auxiliary systems such as **infotainment**.
2. Second Press - Initiates the auxiliary system, turning on low voltage components and dashboard warning lights.
3. Third Press - To ensure the driver is ready to start the vehicle, they must place their foot on the brake while pressing the button a third time. This puts the vehicle in ready mode, making it fully operational and ready to drive.

Figure 2.20 Push Button Start

Hybrid and Electric Vehicles

Driver Display

While, many hybrid and electric vehicles may still have **analogue dashboard** dials or components, they typically feature electronic animations for power distribution, as well as calculations for drive battery charge status and remaining range. The dashboard also includes warning lights for malfunctions, driver information, and an indicator showing when the vehicle is ready to drive.

Figure 2.21 Drivers Display

Smart key - a smart key can unlock, lock, or start a vehicle without inserting a key into the ignition, as long as the driver has the key fob with them.

Infotainment - a combination of information and entertainment.

Analogue dashboard - a drivers display that uses dials and needles instead of electronic digital information.

Auxiliary 12v Battery

The auxiliary system in hybrid and electric vehicles, which includes lights, wipers, and infotainment, operates on a low-voltage system of 12 volts, similar to a conventional vehicle. This system is often powered by a traditional 12 volt battery, which can be either lead acid or lithium-ion. This battery is usually smaller in physical size and capacity compared to those in traditional internal combustion engine vehicles. Once the high-voltage system is activated, the auxiliary system gets its electricity from the high-voltage battery through a step-down transformer known as a DC-to-DC converter.

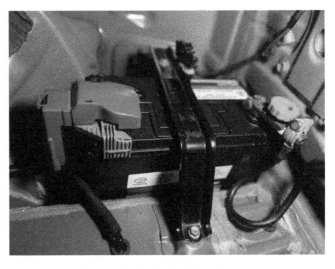

Figure 2.22 12 Volt Auxiliary Battery

Hybrid and Electric Vehicles

Batteries and Types

Hybrid and electric vehicles use high-voltage batteries for energy storage, which power their propulsion systems. However, due to the risks associated with high-voltage batteries, auxiliary systems (which control electrics like headlights, wipers, windows, entertainment, etc.) are usually powered by a low-voltage source. This setup is similar to that of traditional petrol and diesel engine vehicles.

Lead acid batteries, like those found in internal combustion engine vehicles, are typically used to support the low-voltage auxiliary system. However, some manufacturers are starting to use lithium-ion batteries for this purpose. High-voltage batteries are primarily of nickel-metal hydride or lithium-ion design. These types of batteries have a higher energy density (more energy for their size and weight) than lead acid batteries.

Nickel-metal hydride batteries have a higher energy density than lead acid ones, but they're usually only used in hybrid vehicles where the main propulsion comes from an internal combustion engine supplemented by an electric motor. This is because their energy density isn't considered sufficient for fully electric vehicles.

Lithium-ion batteries, on the other hand, have enough energy density to power fully electric vehicles, making them the most viable option currently available. Other battery chemistries are being developed and may replace these designs in future vehicles.

Lead Acid Batteries
Advantages:
- They are cost-effective and widely available.
- They do not develop a memory effect when discharged and recharged.

Disadvantages:
- They can be damaged if fully discharged.
- They have a low energy density compared to other battery types.
- Lead, a component of these batteries, is toxic.

Nickel Metal Hydride Batteries
Advantages:
- They can be recharged quickly, and their materials are readily available.
- They have a relatively high efficiency and are often considered more environmentally friendly than other high-voltage battery types.

Disadvantages:
- They contain a very strong caustic alkaline electrolyte.
- If they catch fire, they can produce toxic fumes, including oxides of nickel, cobalt, aluminium, manganese, lanthanum, cerium, neodymium, and praseodymium.

Lithium-Ion Batteries
Advantages:
- They produce a high nominal voltage per cell, resulting in a large energy density for their size, volume, and weight. This makes them highly suitable for use as the main propulsion battery in electric vehicles.

Disadvantages:
- Lithium is a highly reactive chemical. If short-circuited or overheated, the battery may undergo thermal runaway, which can result in uncontrolled combustion and potential explosion.
- A by-product of thermal runaway is hydrogen, which is also highly flammable. If the battery catches fire, it can be very difficult to control or extinguish.

Hybrid and Electric Vehicles

Regardless of the battery chemistry used, the risk of electrocution from high-voltage is high. Therefore, all batteries should be handled with extreme caution.

As batteries are a form of electrochemical power storage, they are always active and can never be fully shut down.

HV Isolator MSD and HVIL

Hybrid and electric vehicles often include a removable high-voltage isolator plug (MSD). This plug is used to cut off the high-voltage from the battery to the rest of the vehicle during maintenance or repairs, ensuring safety around high-voltage components.

> Some isolators break the circuit within the battery itself to prevent current flow.
> Others interrupt one of the high-voltage wires leading from the battery.

This plug should only be removed by trained operators wearing appropriate personal protective equipment (PPE).

Some manufacturers use a different component known as a high-voltage interlock loop (HVIL), which interrupts the 12 volt system controlling the connection of the high-voltage battery to the rest of the circuit, instead of providing a high-voltage isolator plug.

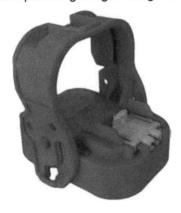

Figure 2.23 An MSD   **Figure 2.24** An HVIL

The high-voltage isolator plug can be referred to by several different names, including a service plug grip and a Maintenance or Manual Service Disconnect (MSD).

High-voltage interlock loops (HVIL) are often used in the construction of high-voltage components as an extra safety measure and are often incorporated in the design of an MSD. They activate if someone begins to dismantle the high-voltage system without first isolating the drive battery. However, these should not be solely relied upon for protection. It's crucial to follow all manufacturer procedures for shutdown and isolation, and to wear appropriate high-voltage Personal Protective Equipment (PPE).

Hybrid and Electric Vehicles

Inverters

Electric vehicles typically use alternating current (AC) for their drive motors, while the energy stored in batteries is in direct current (DC). To power the drive motors, the DC from the battery needs to be converted into AC. This conversion is done by a unit called an inverter, which is placed between the battery and electric drive motors. The inverter contains electronic components called transistors that carry out this conversion process, known as inversion.

When electricity is created by the motor-generators, it's in the form of alternating current (AC). To charge the batteries, this AC must be converted back to direct current (DC). This process, known as rectification, is also done inside the inverter.

Figure 2.25 An Inverter

DC to DC Converter

Hybrid and electric vehicles use a low-voltage 12 volt system to power the vehicle's auxiliary electrics. A 12 volt low-voltage battery is provided for this purpose. However, once the vehicle is in ready mode, the high-voltage battery supplements the low-voltage supply through a step-down electrical transformer known as a DC-to-DC converter. This converter provides energy to operate the 12 volt auxiliary system and keeps the low-voltage 12 volt battery charged.

Figure 2.26 A DC-to-DC Converter

High-Voltage Cables

High-voltage energy in vehicles is kept separate from the rest of the vehicle and its chassis/frame through the use of fully insulated cables. These cables are usually brightly coloured, often orange, to indicate their high-voltage potential. However, the colour is not fully standardised by law, so caution is necessary around any brightly coloured wiring in a hybrid or electric vehicle. Regardless of their size, orange coloured wires or electrical plugs could potentially carry high-voltage and pose a risk.

Figure 2.27 High-Voltage Cables

Battery Management Unit BMU

To oversee and manage the high and low voltage battery and systems, a computer known as the Battery Management Unit (BMU) is needed. The BMU's location can vary, but it's often placed near the high-voltage battery.

Figure 2.28 A Battery Management Unit

Hybrid and Electric Vehicles

Traction Motors

Powerful motors, often hidden within the vehicle's transmission system, are used for electrical propulsion. When these motors receive electricity, they rotate mechanically. Most electric drive motors are designed to use alternating current from a three-phase power supply. This creates a moving magnetic field that turns a component called the rotor, which is connected to the final drive that propels the wheels. When an electric motor rotates mechanically, it generates electricity that can be used to charge the high-voltage battery. Since these processes are combined in one unit, it is often referred to as a motor-generator.

Figure 2.29 A Traction Motor

Generators

When a motor is mechanically turned, it transforms into a generator, producing electricity. In hybrid vehicles, this process allows the engine to generate electricity for the high-voltage battery. Both hybrid and fully electric vehicles can harness the rotational energy from the drive motor, converting it into electricity. This energy exchange slows down the vehicle, a phenomenon known as regenerative braking. The energy that would typically be lost as heat in traditional friction brakes can instead be used to supplement the charge of the high-voltage battery.

Figure 2.30 A Generator

Full hybrids, which can't be charged via an electrical outlet and rely on an engine-driven generator to recharge their high-voltage battery, are often called 'self-charging hybrids'. This term can be somewhat misleading as it might suggest that the electrical energy is free or highly efficient. However, the electricity in these vehicles is generated from a petrol or diesel internal combustion engine, and additional loads can increase fuel consumption and emissions. Even if the electricity from an outlet comes from an oil or gas-fired power station, it's still more efficient than using the vehicle's internal combustion engine. Therefore, plug-in hybrids and electric vehicles generally produce fewer environmental emissions than a full hybrid.

On-Board Chargers OBC

Plug-in hybrids or fully electric vehicles are primarily charged from an electrical outlet. Both domestic and commercial electricity is supplied as alternating current (AC). However, the high-voltage battery systems that power these vehicles use chemically stored energy in the form of direct current (DC). Therefore, when charging from the mains, a component is needed to convert the AC supply to DC in order to charge the battery. This is the role of the onboard charger. This component, carried by the electric vehicle, performs the conversion from AC to DC and facilitates communication between the vehicle and the external charging unit, known as Electric Vehicle Supply Equipment (EVSE). This two-way communication ensures that both the charging unit and the vehicle can safely and efficiently deliver and receive the correct amount of electricity.

Figure 2.31 An On-Board Charger

Hybrid and Electric Vehicles

Air Conditioning (AC) Compressor

In traditional vehicles with internal combustion engines, the air conditioning and climate control system's refrigerant compressor (or pump) is mechanically driven by a belt connected to the engine's crankshaft. However, hybrid and fully electric vehicles may not have the ability to mechanically drive the air conditioning compressor. As a result, most of these systems use a high-voltage electrically driven air conditioning compressor.

Even in a hybrid vehicle with an internal combustion engine, the engine may not always be running, which would affect the functioning of the air conditioning if a mechanical pump was used during an auto-stop phase. To ensure efficiency, the air conditioning compressor is powered by a **three-phase** electric motor that uses the same voltage as the high-voltage system.

Figure 2.32 An Air Conditioning Compressor

 The air conditioning and climate control are always active while the system is operating. Due to the high-voltage involved, caution must be exercised when working on or around the air conditioning system

Although the refrigerant cycle is similar to that used in vehicles with internal combustion engines because the compressor uses high-voltage, a special non-electrically conductive lubricating oil must be used. If incorrect oil is used or if the system is cross-contaminated from another vehicle, there could be a risk of high-voltage insulation breach. This could potentially electrify the vehicle chassis/frame with high-voltage.

PTC Heating

Heating the vehicle cabin is an essential function of the heating, ventilation, and air conditioning (HVAC) system in any vehicle. Traditionally, heat absorbed from the engine into its cooling system was directed through a small radiator inside the car, known as a heater matrix. Ventilation air blown through this heater matrix could be used to regulate the temperature and heat the vehicle cabin.

However, plug-in hybrid electric vehicles and fully electric vehicles may not have the ability to use engine coolant as a heat source inside the vehicle. Therefore, an alternative method is needed. **Positive Temperature Coefficient (PTC)** heaters, a type of electric heating device, provide a solution. They pass electricity through a heating element, which can either heat a liquid (used in a similar manner to those in a heater matrix) or provide an all-electric heat source. Air can be blown through this to create warmth for the passengers and occupants of the vehicle.

Figure 2.33 A PTC Heater

 **Three-Phase** - an electrical system that uses three separate alternating currents of the same voltage.

Positive Temperature Coefficient (PTC) - a characteristic where electrical resistance increases as temperature rises.

Hybrid and Electric Vehicles

Charging an Electric Vehicle

When discussing the operation of charging an electric vehicle from mains electricity, two key terms are used: Mode and Type.

Mode refers to the charging method, which includes whether it is AC or DC, and whether it is slow, fast, or rapid charging.

Table 2.7 describes the Modes of charging.

Table 2.7 Charging modes	
Mode	**Description**
Mode 1	Now considered mostly obsolete, this is a simple extension lead that plugs into a domestic mains or commercial electricity supply with no intelligence or safety other than the fuse in the plug or a circuit breaker.
Mode 2	An intelligent extension lead supplied with an in-cable charging box (ICCB), which can communicate with the vehicle's onboard charger. This uses single-phase domestic electricity only (Slow).
Mode 3	Dedicated AC charging that can be either single or three-phase, depending on location and electricity supply (Fast).
Mode 4	Specialist DC rapid charging. In this system, alternating current is converted to direct current in the Electric Vehicle Supply Equipment (EVSE), which can then bypass the vehicle's onboard charger directly to the battery (Rapid).

Type refers to the shape of the plug used to connect the electric vehicle to the Electric Vehicle Supply Equipment (EVSE).
Table 2.8 describes the charging plug types.

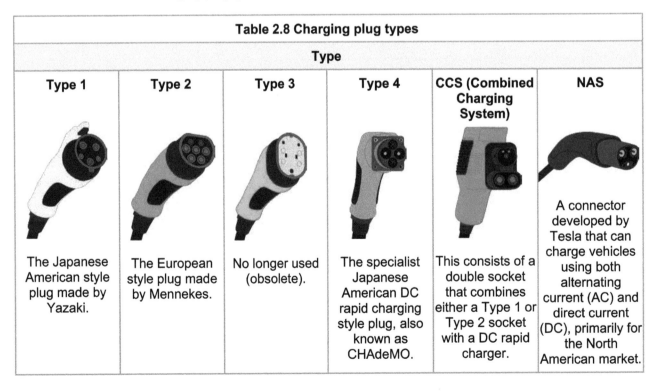

Table 2.8 Charging plug types					
Type					
Type 1	Type 2	Type 3	Type 4	CCS (Combined Charging System)	NAS
The Japanese American style plug made by Yazaki.	The European style plug made by Mennekes.	No longer used (obsolete).	The specialist Japanese American DC rapid charging style plug, also known as CHAdeMO.	This consists of a double socket that combines either a Type 1 or Type 2 socket with a DC rapid charger.	A connector developed by Tesla that can charge vehicles using both alternating current (AC) and direct current (DC), primarily for the North American market.

Hybrid and Electric Vehicles

Regardless of mode or type, always consult and follow manufacturers' instructions when charging an electric vehicle.

When using a public charger, if the parking space is occupied by a petrol or diesel engine vehicle that prevents the electric vehicle from accessing it, this is often referred to as 'iced'. The term 'iced' relates to the charger being blocked by an internal combustion engine (ICE).

Alternative Propulsion Comparisons

Table 2.9 shows comparisons in the approximate efficiency for different fuel and vehicle drive types.

Table 2.9 System drive types and efficiency	
Fuel/Drive system type	**Approximate efficiency**
Steam produced by coal	In practice, a steam engine exhausting the steam to atmosphere will typically have an efficiency (including the boiler) in the range of 1-10%; this means that it is around 90% inefficient. With the addition of a condenser and multiple expansion, this figure can be improved to 25% or better.
Petrol	The efficiency of a petrol-powered vehicle is surprisingly low. All of the heat that comes out as exhaust or goes into the radiator is wasted energy; the engine also uses a lot of energy turning the various pumps, fans, and generators that keep it going. So, the overall efficiency of a petrol engine is about 20%. That is, only about 20% of the thermal energy content of the petrol is converted into mechanical work.
Diesel	Diesels are able to reach an efficiency of about 40% in the engine speed range of idle to about 1800 rpm. After this speed, efficiency begins to decline due to air-pumping losses within the engine.
Hybrid	Depending on the hybrid drive type: ➤ Series ➤ Parallel ➤ Power split ➤ Dual Efficiency values will vary. If you combine the efficiency of a petrol engine of around 20%, with the efficiency of electric motors of around 90%, then in theory you should get an overall efficiency of about 70%. Unfortunately, due to construction and design, heat losses will reduce this figure to around 55% to 60%, putting them in the same sort of efficiency as a modern common-rail diesel. These figures though do not take into account the large reduction in emissions produced by hybrids when compared to diesel engines.

Hybrid and Electric Vehicles

Table 2.9 System drive types and efficiency

Plug in electric	A battery-powered electric vehicle has a fairly high efficiency. The battery is about 90% efficient (most batteries generate some heat, or require heating), and the electric motor/inverter is about 80% efficient. This gives an overall efficiency of about 72%. But that is not the whole story, the electricity used to power the vehicle had to be generated somewhere. If it was generated at a power plant that used a combustion process (rather than nuclear, hydroelectric, solar or wind), then only about 40 percent of the fuel required by the power plant was converted into electricity. The process of charging the vehicle requires the conversion of alternating current (AC) power to direct current (DC) power. This process has an efficiency of about 90 percent. So, if we look at the whole cycle, the efficiency of an electric vehicle is 72% for the vehicle, 40% for the power plant and 90% for charging the vehicle. That gives an overall efficiency of 26%. The overall efficiency varies considerably depending on what sort of power plant is used. If the electricity for the vehicle is generated by a hydroelectric plant for instance, then it is basically free (no fuel was burned to generate the electricity), and the efficiency of the electric vehicle is about 65%.
Solar	Because photovoltaic cells (solar panels) are unable to convert all of the energy in the electromagnetic spectrum produced by sunlight, they are only currently around 10% efficient. This means that a 1 square meter solar panel that is capable of receiving 1 kilowatt of energy from the sun can only convert this into 100 watts of energy. This is around 1.3 horsepower (Hp). This very low power output makes solar panels unsuitable for powering most vehicles, but they can be used to extend the range of plug-in electric vehicles or be employed in the electrolysis processes used to create hydrogen or ammonia which can then be used in vehicle propulsion.
Hydrogen powered ICE	When compared to a standard internal combustion engine running on petrol, a hydrogen engine has the potential to produce a greater power output because of the higher compression ratios that can be used. Unfortunately, due to the space required inside the engine cylinder by the volume of hydrogen in its gaseous state, air-fuel ratios tend to run extremely weak leading to a low efficiency overall of around 25%.
Hydrogen fuel cell	If the fuel cell is powered with pure hydrogen, it has the potential to be up to 80% efficient. That is, it converts 80% of the energy content of the hydrogen into electrical energy. However, we still need to convert the electrical energy into mechanical work. This is accomplished by the electric motor and inverter. A reasonable number for the efficiency of the motor/inverter is about 80%. So we have 80% efficiency in generating electricity, and 80% efficiency converting it to mechanical power. That gives an overall efficiency of about 64%.
LPG, CNG and biogas	Although liquefied-petroleum gas LPG is generally cheaper to buy than the equivalent quantity of petrol or diesel, it is less fuel-efficient than its alternatives. A vehicle fitted with an LPG system will use more fuel than an equivalent vehicle fitted with a petrol engine and will therefore get fewer miles per gallon (MPG). A factor of approximately a 15% efficiency reduction should be taken into consideration for any fuel economy calculations when compared to the equivalent petrol.

Hybrid and Electric Vehicles

Table 2.9 System drive types and efficiency	
Ammonia green NH$_3$	When compared with petrol, ammonia green NH$_3$ has a very similar overall energy content. Due to the way it is manufactured, only around 70% of the energy put into the production process is converted into hydrogen and therefore, when burnt in a standard internal-combustion engine, its efficiency will be approximately 28%.
Bioalcohol/Ethanol	Bio-alcohol or ethanol produced from crops is approximately 34% less efficient than the equivalent petrol. A standard engine running on ethanol would therefore not produce the same fuel economy as a comparable petrol engine. As ethanol has a higher octane rating than petrol, it does however, increase the ability of engine designers to raise compression ratios or use forced-air induction and therefore regain some of its overall efficiency.

Summary

This chapter has described:

- Why we need electric vehicles.
- Different types of alternative propulsion used in vehicles.
- Different types of hybrids and electrically propelled vehicles.
- Electric vehicle systems, components and operation.
- Charging vehicles from mains electricity.

Electric Vehicle Hazards and Maintenance

Chapter 3 Electric Vehicle Hazards and Maintenance

This chapter introduces you to the hazards associated with electric vehicles. You will be provided with information describing how to work on and around these vehicle types while performing maintenance that does not directly involve the high-voltage systems. The chapter also contains advice on safety strategies and procedures that should be implemented to reduce the possibility of accidents, and the care that needs to be taken when interacting with electric vehicles.

By definition, an accident is something bad that happens that is not expected or intended, which often damages something or injures someone. Accidents happen by chance without anyone planning them and will have consequences for anyone involved. Therefore, it is important that precautions are taken which reduce the risks associated with vehicle recovery, maintenance, and repair.

By reading this chapter, you will gain a better understanding of the potential hazards, and actions which could be implemented as good working practice. You will also learn how to interact with these vehicles safely and efficiently, whether you are a driver, a passenger, a technician, a member of the automotive industry, or a first responder. This chapter is important for those who conduct routine maintenance or provide roadside recovery following breakdown or accidents.

Contents

- ❖ Information sources **Page 73**
- ❖ Safety systems **Page 73**
- ❖ Hazard management **Page 76**
- ❖ Legislation and regulations **Page 80**
- ❖ Fire extinguishers and suppressants **Page 82**
- ❖ Personal Protective Equipment (PPE) and first aid **Page 85**
- ❖ Electric vehicle hazards and maintenance **Page 92**

The automotive industry is a high-risk environment, especially when dealing with electrical systems. The hazards of electricity are well-known, but they can be easily ignored due to its invisible nature. This can lead to complacency unless the fundamental operation of electric vehicles is understood. Even with this understanding, caution is necessary. Do not rely on any safety systems designed for protection; instead, take precautions to minimise the risk of injury or death. Always evaluate the risks associated with any activity and implement measures to eliminate or reduce the hazards involved in any task that involves hybrid or electric vehicles.

Additional risks associated with working on, or around electric vehicles may include:

- ➤ Electrocution
- ➤ Strong magnetic fields
- ➤ Falling from height
- ➤ Injuries caused by incorrect manual handling techniques
- ➤ Chemicals
- ➤ Gases or fumes
- ➤ Hybrid engine systems starting unexpectedly
- ➤ The silent movement of electric vehicles while in use

Never attempt to work on a high-voltage electrical system unless you have received adequate training.

Electric Vehicle Hazards and Maintenance

Information Sources

Servicing and maintaining hybrid and electric vehicles requires a reliable source of technical information and data. Before starting and during the maintenance and repair process, it's crucial to have comprehensive information about these systems.

Potential sources of information may include:

Table 3.1 Possible information sources	
Verbal information from the driver	Vehicle identification numbers
Service and repair history	Warranty information
Vehicle handbook	Technical data manuals
Workshop manuals/Wiring diagrams	Safety recall sheets
Manufacturer specific information or data sheets	Information bulletins
Technical helplines	Advice from other technicians/colleagues
Internet	Parts suppliers/catalogues
Jobcards	Diagnostic trouble codes
Oscilloscope waveforms	On vehicle warning labels/stickers
On vehicle displays	Temperature readings

Always compare the outcomes of any inspection or maintenance with appropriate data sources. Regardless of the information or data source you choose, it's crucial to assess its usefulness and reliability for your maintenance and repair work.

If you need to replace any electrical or electronic components, always ensure that the quality meets the original equipment manufacturer (OEM) specifications. If the vehicle is under warranty, using inferior parts or making deliberate modifications might invalidate the warranty. Additionally, fitting parts of inferior quality could affect vehicle performance and safety. You should only replace electrical components if the parts comply with the legal requirements for road use.

Safety Systems

Safety Signs

Signs are commonly used in the workplace to promote health and safety by providing instructions or warnings. These signs typically include images to help convey their message. They follow certain formats.

Red signs signal a **prohibition**, indicating actions that are not allowed. They usually feature a symbol crossed out by a red diagonal line.

Figure 3.1 A Prohibition Sign

Blue signs represent **mandatory** instructions, signifying actions that must be carried out.

Figure 3.2 A Mandatory Sign

Electric Vehicle Hazards and Maintenance

Yellow signs serve as warnings, indicating hazards or dangers.

Figure 3.3 A Warning Sign

Green signs denote safe conditions or useful information. They are often used to guide directions in emergencies.

Figure 3.4 A Safety Sign

Prohibition - Actions that are not allowed.

Mandatory - Actions that are required.

Warning labels are typically used where there may be health or environmental risks from chemicals or substances. They provide necessary information for safe handling.

Figure 3.5 Warning Labels

Many high-voltage systems and components found on electric vehicles will carry specific safety information and warnings, often on labels. However, this may not always be the case. Care must be taken when working on or around any potential high-voltage system.

Working Perspective and Hazard Management

Working with vehicles that use electric propulsion systems can present unique challenges. These challenges can vary based on your role and the specific situation.

You might be involved in various ways, including:

- Being part of the emergency services like police, fire department, ambulance, or coastguard.
- Working in the motor industry as a recovery operator, vehicle technician, MET, vehicle body/paint specialist, or vehicle recycler.
- Being part of a training program as a lecturer, student, apprentice, workshop instructor, assessor, or mentor.

Electric Vehicle Hazards and Maintenance

The following table describes an example scenario relating to hybrid and electric vehicles and lists a set of suggested precautions or actions that could be undertaken. These actions are not in any particular order or restricted to an individual profession, perspective, or approach.

Using the information provided in this chapter and the knowledge of your own role or position, create your own list of actions that you would take in different situations.

Table 3.2 Potential scenarios

Example scenario	Potential actions
➢ You're about to work on a hybrid or electric vehicle, either roadside or in a workshop. ➢ The task might be small, with potential risks like interacting with high-voltage systems or unexpected engine starts. ➢ The task could be significant, involving high risks like dealing with a severely damaged vehicle or one that's been submerged in water.	➢ Assess the area for possible hazards. ➢ Avoid working near the strong magnetic fields from electric vehicles if you have life-sustaining electronic equipment. ➢ Wear high-visibility clothes. ➢ Make sure others know what's happening (avoid working alone if possible). ➢ Carry out a risk assessment. ➢ Set up a safety zone around the vehicle with signs and barriers. ➢ Keep a suitable fire extinguisher nearby. ➢ Remove any potential sources of ignition. ➢ Turn off the ignition and keep the smart key out of range. ➢ Secure the vehicle to stop unexpected movement. ➢ Wear any necessary high-voltage or chemical-resistant personal protective equipment (PPE). ➢ Let any water drain naturally from the vehicle (if it's been submerged). ➢ Remove electrical system fuses. ➢ Isolate the ignition, fuel injection, and **SRS** airbags. ➢ Follow manufacturer's procedures to isolate the high-voltage system via the **MSD** or **HVIL**. ➢ Check that system voltage has dropped as close to zero as possible from the HV battery and at the capacitors. ➢ Prevent accidental reconnection of the high-voltage system.

MSD - Maintenance Service Disconnect or Manual Service Disconnect, also known as the isolator plug or service plug grip. It is a high-voltage plug that can be removed before maintenance to isolate the battery from the rest of the vehicle. (High-voltage PPE must be worn for removal and reconnection).

HVIL - High-Voltage Interlock Loop, a component designed to interrupt the low voltage supply to the battery contactors or system main relays (SMR).

SRS - Supplementary Restraint Systems, are additional passenger safety systems that include components such as air bags and pyrotechnic pre-tensioners.

Waterproofing and Submersion

Electric vehicles need to waterproof their high-voltage systems, according to the ECE R 100 regulations. This primarily involves two main areas and outlines the minimum protection against high-voltage isolation resistance, direct and indirect contact.

Electric Vehicle Hazards and Maintenance

Washing

This test mimics a typical car wash but excludes high-pressure cleaning or washing the underbody. The focus is on the vehicle's border lines, which are the seals between two parts like flaps, glass seals, opening part outlines, front grille outlines, and lamp seals. Water is sprayed from all directions for a certain duration to ensure that high-voltage components are shielded from direct spray and splashing.

Driving Through Standing Water

This test replicates driving through standing water. Vehicles are driven in a wade pool with a water depth of 10 cm, covering a distance of 500 m at a speed of 20 km/h, for about 1.5 minutes. If the wade pool is less than 500 m long, the vehicle is driven through it multiple times.

Electric vehicles are waterproofed to a certain standard when new, meaning they can be safely operated and washed using conventional methods or in a car wash. However, pressure washing should be avoided, especially around vulnerable or high-voltage components.

The high-voltage battery in many electric vehicles is typically located low down, often forming part of the vehicle frame or floor pan. While battery casings are designed to be sealed against water ingress, this protection may deteriorate over time due to aging, wear, and road damage. Manufacturers usually provide a maximum wading depth for driving through standing water in the vehicle literature, but this is often overlooked. It's important to locate this information and make the vehicle operator aware of it.

In situations where a vehicle has been submerged, such as in a flood, it should be approached with extreme caution. Safety measures should be put in place following the manufacturer's recommended procedures and precautions.

Hazard Management

Hazard management is a vital component to be considered in any approach to dealing with an electric vehicle. This must always be the first action, especially when attending an accident or roadside recovery. There are many safety systems included in the design of electric vehicles, but often in an emergency or breakdown situation, these may be disabled or compromised.

The following section regarding accidents is designed to discuss access to and the recovery of vehicles, but it does not address any consequences for drivers, operators, or passengers involved.
If available, manufacturer emergency response guides should be referred to for vehicle-specific procedures and sequences.

Accidents

Accidents pose a significant risk due to exposed high-voltage components. The risk of electrocution increases, along with other hazards such as leaking electrolyte, strong magnetic fields, and an elevated possibility of fire or explosion.

Following an accident, vehicle safety systems are designed to shut down and isolate the high-voltage battery from the rest of the vehicle, if possible. However, other system components have the potential to store or generate electricity. Therefore, the situation must always be assessed, and appropriate precautions must be taken.
Every automotive accident will differ in cause, severity, and outcome. Every action used to secure the scene and render the vehicle safe should be viewed and used depending on its merits relating to the situation.

Electric Vehicle Hazards and Maintenance

Lack of engine noise does not mean that the vehicle is switched off. Silent movement or instant restart is a feature associated with hybrid and electric vehicles. This capability is available until the vehicle has been fully shut down. Therefore, always take reasonable precautions to prevent accidental restart or movement and wear appropriate personal protective equipment (PPE).

Capacitors are used in the design and operation of vehicle drive systems. They are used by the motor generators as a quick access storage device for electrical energy. Charged from the high-voltage battery, the system is designed to empty the capacitors whenever the operator switches off the vehicle. In the event of an accident, the active discharge of the capacitors may not be possible. Manufacturers equip the system with a backup to empty the capacitors using an electrical resistor connected within the circuit. This backup or redundancy system takes time to be effective and bring the stored voltage down to a safe level. As long as the backup capacitor discharge system has not been compromised during the accident, where possible a wait time of around ten minutes should be observed before approaching or coming into contact with a damaged vehicle.

Figure 3.6 High-Voltage Capacitors

The passive discharge time of system capacitors will vary from manufacturer to manufacturer. Therefore, never assume that the system is safe until this can be confirmed using appropriate and calibrated electrical test equipment, following the manufacturers' prescribed methods and procedures.

When attending an accident, an initial assessment of the location and situation should be conducted. Any external hazards, including chemical spills, sharp objects, and sources of ignition, for example, should be dealt with, if possible, before attending to the vehicle.

Immobilisation

It is possible that an electric or hybrid vehicle may move or roll unexpectedly, so it should be immobilised by blocking or chocking the wheels, etc. Place the vehicle in park, switch off and remove any smart key beyond its range of operation.

Figure 3.7 Chocking Wheels

Electric Vehicle Hazards and Maintenance

First Responder Loops

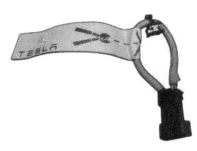

Some manufacturers supply a first responder loop, which is an area of electrical wiring that can be cut in the event of an emergency to isolate and disable the low-voltage auxiliary system, especially related to the battery contactors and air bag systems. Often found in emergency response guides, this section of wiring will be highlighted with a label showing the emergency services where to cut. Be aware that not every high-voltage component is labelled, so always wear appropriate personal protective equipment (PPE). First responder loops are often designed to be cut in two places, removing a section that should reduce the possibility of wiring bare ends touching and reconnecting the circuit by mistake.

Figure 3.8 A First Responder Loop

High-Voltage Isolation

Regardless of whether a first responder loop has been disconnected, the high-voltage system of the vehicle may still be energised and therefore dangerous. Always treat any components, cables, or wires, often highlighted with the colour orange, as if there is high-voltage in them. Follow the manufacturers' recommended procedures for shutdown, isolation, and absence of voltage confirmation. Due to the chemical nature of batteries, it is impossible to fully discharge or shut them down and therefore additional precautions must be taken so that the battery pack is not damaged or compromised during any recovery procedure.

Be aware that the restraint control module (RCM) and supplementary restraint systems (SRS) may have an internal energy reserve that allows them to remain powered for some time after the 12 volt system is disconnected.

Airbag and pretensioner systems may have stored gas inflators. Rescuers should never cut or crush inflation cylinders, as this may cause catastrophic failure leading to injury or death.

Lifting Points

The high-voltage battery of many vehicles is mounted below the floor pan and forms part of the undercarriage. Specific stabilisation and lifting or jacking points are available and must be used to prevent damage to the battery casing and internal battery components.

Roadside Recovery - Including Towing

The drive motors of many electric vehicles are often permanently connected to the driven axle. This means that if a vehicle is pushed or towed, the rotation of the drive wheels can cause the motors to generate electricity. If not managed, this electricity may cause damage, injury, or overheating of the systems. In extreme cases, overheating could cause the surrounding components to ignite, potentially igniting the high-voltage battery as well. In general, never move or tow an electric vehicle with the driven wheels on the ground, unless you are pushing it a short distance at very low speed to clear a roadway. Driven wheels should be supported on roller dollies or correctly lifted and suspended above the ground during any movement of the vehicle.

Figure 3.9 Recovering an Electric Vehicle (Front Wheel Drive)

Electric Vehicle Hazards and Maintenance

 Some vehicles might have a transport mode that can be engaged if the vehicle is suitably powered and safe to do so. This may allow the vehicle to be pushed or moved a short distance to safety.

Jump Starting

The main purpose of the low-voltage auxiliary battery is to allow the engagement of the system main relays (SMR) or contactors in order to power up the high-voltage system. Once powered, the auxiliary load and operation are managed via the **DC-to-DC converter**. This means that the auxiliary battery is normally a small low-capacity unit, as it only has to supply energy to maintain essential systems that require power when the vehicle is not in use, such as clocks, radios, alarms, immobilisers, and central locking, for example. As a result, any **parasitic drain** that is maintained for a period of time may cause the auxiliary battery to fall below the required voltage necessary to engage the system main relay or contactors. This will provide a non-start situation, similar to those found in traditional internal combustion engine vehicles with a flat battery. If this occurs, it may be necessary to boost the voltage of the low-voltage auxiliary circuits in order to engage the high-voltage system and power up the vehicle, placing it in ready mode.

It is possible to jump start a vehicle in a manner similar to that traditionally used for other vehicles, but several precautions need to be observed.

Many manufacturers place the 12 volt auxiliary battery in a location that is difficult to access. As a result, specific jump points in maintenance access areas, such as the fuse box, may be provided. It is important to follow any manufacturer-specific guidance when connecting a jump pack or jump leads to these points, ensuring correct **polarity** at all times.

If the 12 volt battery is accessible, it may be possible to connect or jump start the vehicle at this point but be aware of any sensitive battery measuring devices that may be attached and damaged during any jump start procedure.

Unlike a traditional internal combustion engine vehicle, a large amount of current is not required to crank the engine via a starter motor. This means that the voltage potential is the important component for raising the level enough to engage system main relays or contactors. Once a suitable power source has been correctly connected, the vehicle should be placed in ready mode, at which point the DC-to-DC converter will take over and manage the low-voltage auxiliary system and recharging of the low-voltage auxiliary battery.

Take care when disconnecting any crocodile clips from the auxiliary jump points or battery to ensure that short circuit or arcing does not occur.

Figure 3.10 Jump Starting

Electric Vehicle Hazards and Maintenance

DC-to-DC converter - a step-down transformer which converts high-voltage direct current (DC) from the drive battery to low-voltage direct current (DC) to power the auxiliary system and charge the 12 volt battery.

Parasitic drain - a term that describes the situation when the vehicle battery continues to lose charge even when it is turned off or not in use.

Polarity - the property of having two distinct ends or poles, each with a different electric charge. In an electric circuit, polarity determines the direction of the current flow.

Legislation and Regulations

The construction, use, operation, and maintenance of hybrid and electric vehicles are governed by legal requirements, regulations, and standards. These vary depending on the country of origin, intended market, and geographical location. Examples of legislation, regulations, and standards that may be applicable can be found in the appendix. However, ensure that you comply with the geographical requirements of the country in question.

Risk Assessments

Working in the motor industry and with high-voltage vehicle electrical systems or fuel used in alternative propulsion is extremely hazardous. There are many dangers that could result in an accident causing injury or even death.

The management of these **hazards** is key to reducing the **risks** involved while working on these systems. Not all hazards can be removed, but they can be identified, and measures put in place to reduce the dangers that they pose. This is the purpose of a risk assessment.

A risk assessment is an important step in protecting workers and businesses and is necessary in order to comply with the Management of Health and Safety at Work. It is designed to focus on the risks that really matter in the workplace – the ones with the potential to cause real harm. In many cases, straightforward measures can control risks. The law does not expect you to eliminate all risk, but you are required to protect people as far as is **reasonably practicable**.

Hazard - something that has the potential to cause harm or damage.

Risk - the likelihood of the harm or damage actually occurring.

Reasonably practicable - can be carried out without excessive effort or expense.

A risk assessment is simply a careful examination of what, in your work, could cause harm to people. It allows you to weigh up whether you have taken enough precautions or should do more to prevent harm.

There are five main steps to risk assessment:

1. Identify the hazards.
2. Decide who might be harmed and how.
3. Evaluate the risks and decide on precautions.
4. Record your findings and implement them.
5. Review your assessment and update if necessary.

Electric Vehicle Hazards and Maintenance

A dynamic risk assessment is a way of assessing and managing the risks and hazards in a changing or high-risk environment.
It involves continuously observing and analysing the situation and making quick decisions to ensure safety.
A dynamic risk assessment is different from a formal risk assessment, which is prepared in advance and recorded. It can complement and fill in any gaps that the formal risk assessment could not predict.
A dynamic risk assessment can be used in circumstances where the work environment is unpredictable, such as emergency, accident, or recovery situations. If the risks are too high or cannot be controlled, the work should be delayed until additional safety measures can be introduced.

COSHH

The legislation that you and your employer must observe when using hazardous substances in the workshop is the Control of Substances Hazardous to Health Regulations (COSHH).

There are eight steps that employers must take to protect employees from hazardous substances.

To comply with COSHH, employers must follow these eight steps:

1. Identify the hazards and assess the risks from hazardous substances.
2. Implement measures to eliminate or reduce the risks.
3. Ensure that exposure is controlled by using appropriate methods and equipment.
4. Check that the control measures are working effectively and regularly maintained.
5. Monitor the exposure levels and health effects of workers.
6. Provide adequate training and information for workers on how to handle hazardous substances safely.
7. Plan for emergencies involving hazardous substances and have suitable procedures in place.
8. Keep records of risk assessments, exposure monitoring and health surveillance.

Workshop Requirements

When planning workshop space or requirements, areas to consider could include:

- ☑ Suitable floor: Marked or delineated areas of work which can easily be identified as high-voltage vehicle work areas.
- ☑ Lifts, ramps, or hoists: They should be suitable for the task being conducted. Two post lifts may have adjustable arms to allow them to be spaced appropriately for jacking points and access to high-voltage battery components. If battery lifting tables are to be used, the floor area between the posts should be flat and unencumbered.
- ☑ Automotive workshop lighting: It should ensure adequate and safe illumination for vehicle repair and maintenance activities. Lighting is an important factor that affects the quality, productivity, and safety of the work environment.
- ☑ Automotive workshop ventilation and extraction: It should ensure adequate and safe removal of vehicle exhaust fumes and other airborne contaminants from the work environment. Vehicle exhaust fumes can irritate the eyes and respiratory tract, and are a risk to health by breathing in. They contain carbon monoxide, a poisonous gas, and diesel fumes, which may increase the risk of lung cancer.

Figure 3.11 An Automotive Workshop

Electric Vehicle Hazards and Maintenance

Disposal of Waste Materials and Environmental Protection

Environmental damage can be caused by contaminating the atmosphere, water supply, or drainage system. Under the Environmental Protection Act (EPA), it is an offence to treat, keep, or dispose of **controlled waste** in a way that is likely to pollute the environment (**ecotoxic**) or harm people. Waste producers must make sure that it is passed only to an authorised person who can transport, recycle, or dispose of it safely. You should have procedures in place for working with and disposing of any material that has the potential to harm the environment.

Figure 3.12 Ecotoxic Symbol

Controlled waste - any waste that cannot be disposed of to landfill, including liquids, asbestos, tyres, and waste that has been decontaminated. There are three types of controlled waste listed under the Environmental Protection Act: household, industrial and commercial waste.

Ecotoxic - a substance that is harmful to the environment.

Regulations describe the actions that anyone who produces, imports, keeps, stores, transports, treats, recycles, or disposes of controlled waste must take.

These people must:

- Store the waste safely so that it does not cause pollution or harm anyone.
- Transfer it only to someone who is authorised to take it (such as someone who holds a waste management licence or is a registered waste carrier).
- When passing it on to someone else, provide a written description of the waste and fill in a transfer note.
- Keep these records and provide a copy to the Environment Agency if they ask for one.

Figure 3.13 A Waste Transfer Note

Fire Extinguishers and Suppressants

Fire safety in the workplace, especially when dealing with automotive related fires, requires careful consideration.

A separate risk assessment should be carried out for fire hazards and safety measures should be implemented accordingly.

Fire Extinguishers

Fire extinguishers should be provided and maintained by the company in case of fire in the workplace. The primary function of a fire extinguisher is to enable you to create an escape route.
You should only attempt to tackle a fire if it is safe to do so and if you have had adequate training.

Electric Vehicle Hazards and Maintenance

Different types of fire extinguishers are available depending on the type of fire to be tackled.

The type of fire is normally classified as follows:

Class A: solids, such as paper, wood, plastic etc.
Class B: flammable liquids, such as petrol, diesel, oil etc.
Class C: flammable gases, such as propane, butane, methane etc.
Class D: metals, such as aluminium, magnesium, titanium etc.
Class E: fires involving electrical apparatus.
Class F: cooking oils and fats, such as vegetable oil, animal fat etc.

Electric vehicles are no more likely to catch fire than their petrol or diesel counterparts, however, once the fire starts, the chemistry inside the batteries can cause a reaction known as **thermal runaway**. Once a battery enters thermal runaway, it is very difficult to control with standard fire extinguishers.

Electric vehicle fires are best dealt with by professional fire services, using large amounts of water discharged over the area of the fire.

Standard fire extinguishers carry a colour-coded label with a description of their contents and a list of the types of fire they are suitable for. You should exercise extreme caution when selecting an appropriate extinguisher to use on a vehicle electrical fire. Liquid-based fire extinguishers could conduct electricity, and this could result in electrocution.

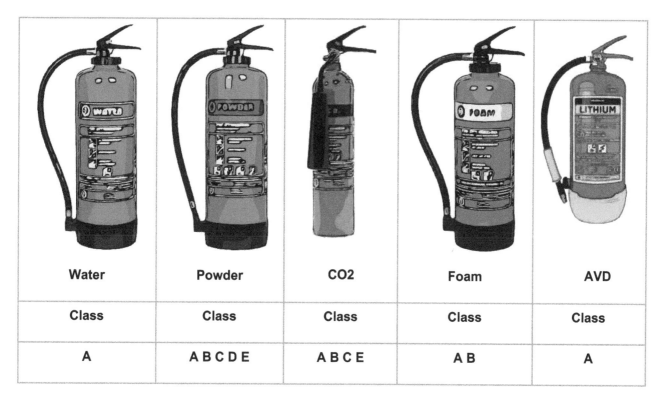

Water	Powder	CO2	Foam	AVD
Class	Class	Class	Class	Class
A	A B C D E	A B C E	A B	A

Specialist fire extinguishers have been developed to help tackle small battery fires containing lithium-ion. Aqueous Vermiculite Dispersion (AVD) is a form of water-based extinguisher, containing a naturally occurring mineral formed from aluminium-iron-magnesium-silicate known as **vermiculite**. The vermiculite is chemically and physically **inert** and has excellent insulation properties.

Electric Vehicle Hazards and Maintenance

How AVD Works

When lithium-ion batteries go into thermal runaway, the process begins with smoke. As the process continues, the battery cells rupture and release hot flammable gases. The AVD fire extinguisher produces a mist that reduces the smoke, cools the flames, and helps to lower the temperature of surrounding battery cells. As the heat evaporates the water, an AVD coating forms over the cells. It creates barriers to heat, oxygen, and electricity.

Thermal runaway is a chemical reaction that occurs within the cells of a battery. It is an exothermic reaction, which produces large amounts of heat and chemical gases very quickly before any flame appears. Exposure to these chemical gases can be highly hazardous.

Chemical gases may include:

- Carbon monoxide
- Carbon dioxide
- Hydrogen fluoride
- Hydrogen chloride
- Sulphur dioxide
- Hydrocarbons such as ethane, methane, butane, and hydrogen cyanide

Smoke Fire Explosion

Granular Fire Suppressants

Granular fire **suppressants** are available that can be used to smother, suppress, and contain fires related to lithium-ion batteries. They consist of small granules that melt when exposed to the high temperatures caused by thermal runaway, turning into glass. As they melt, they absorb a large amount of heat energy from the fire, and the glass forms an impermeable coating that further contains any fire.

Electric Vehicle Fire Blankets

Specialist fire blankets are produced that can be used to help contain the fire from an electric vehicle. Although they do not extinguish the fire, these large, car-sized blankets can smother a burning vehicle, suppressing both fire and heat for a short period of time. This may help provide the opportunity to evacuate the area, allow time for the fire service to arrive, and potentially reduce collateral damage.

Figure 3.14 A Vehicle Fire Blanket

Thermal runaway - a condition in which a system or device produces more heat than it can dissipate, leading to an uncontrollable rise in temperature and potentially destructive consequences.

Vermiculite - a type of mineral that expands when heated and can absorb water.

Inert - not reacting chemically with other substances.

Suppressant - a substance that reduces or stops an undesirable action or condition in the body or the environment.

Electric Vehicle Hazards and Maintenance

Personal Protective Equipment (PPE) and First Aid

Personal protective equipment (PPE) is a vital component of any work that is conducted on or around hybrid and electric vehicles. The main issues that should be considered when selecting PPE for these vehicle types are protection against high-voltage and chemical hazards. Every task will potentially have different requirements and therefore the correct form of PPE should be used. It is essential to remember that PPE is not a substitute for the reduction or removal of hazards from any particular task; it should be used as an added layer to reduce any risks involved where reasonably practicable.

When selecting PPE, make sure that the equipment:

- Is the right PPE for the job – ask for advice if you are not sure.
- Prevents or controls the risk for the job you are doing.
- Fits correctly – it needs to be adjustable, so it fits you properly.
- Is comfortable enough to wear for the length of time you need it.
- Is properly looked after.
- Does not create a new risk, e.g. overheating.
- Does not impair your sight, communication, or movement.
- Is compatible with other PPE worn.
- Does not interfere with the job you are doing.

Table 3.3 describes some personal protective equipment and potential uses.
This list is not exhaustive, and you will need to adapt any recommendations to the specific task you are undertaking.

Table 3.3 Examples of personal protective equipment PPE and potential uses	
PPE	**Recommendations and use**
Overalls or Workwear	Specific workwear or overalls can provide an additional layer of protection between the user and potential hazards. Depending on the materials used, this protection could include containment of loose clothing and the covering of areas of bare skin, resistance to chemicals, or limited fire protection. It is recommended that overalls or clothing avoid metal fastenings when working on or around the high-voltage systems of hybrid and electric vehicles. Metal fastenings may increase the risk of short circuit, leading to electrocution or fire, but also may be affected by the very strong magnetic fields found within the drive systems of electric vehicles.

Electric Vehicle Hazards and Maintenance

Table 3.3 Examples of personal protective equipment PPE and potential uses

Arc Flash/Blast Overalls 	Arc flash/blast overalls are a type of PPE that is designed to protect the wearer from the hazards of arc flash and blast, which are intense bursts of heat, light, and sound that can occur when an electrical fault causes an arc between two conductors. Arc flash and blast can cause severe burns, blindness, hearing loss, shock, and even death. Arc flash/blast overalls are usually made of flame-resistant fabrics that do not ignite, melt, or drip when exposed to high temperatures.
Chemical-resistant Gloves 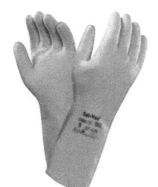	Chemical-resistant gloves are a type of PPE that are designed to protect the hands from harmful substances that can cause skin damage, irritation, burns, or poisoning. Chemical-resistant PPE gloves are made of materials that have different properties and resistance levels against various chemicals. Some examples are: ➤ Natural rubber or latex: This material is flexible, durable and has good resistance to acids, bases, alcohols, and some ketones. However, it is not very resistant to oils, solvents, hydrocarbons, and some organic compounds. It can also cause allergic reactions in some people. ➤ Nitrile: This material is synthetic and has excellent resistance to oils, greases, fuels, solvents, and many organic chemicals. It is also more puncture-resistant than latex and does not cause allergies. ➤ Neoprene: This material is synthetic and has good resistance to oils, greases, alcohols, and acids. It is important to choose the right chemical-resistant PPE gloves for a specific task and consider the following factors: ➤ The type and concentration of the chemicals being handled. ➤ The duration and frequency of exposure. ➤ The temperature and pressure of the work environment. ➤ The size and fit of the gloves. ➤ The compatibility of the gloves with other PPE.

Electric Vehicle Hazards and Maintenance

Table 3.3 Examples of personal protective equipment PPE and potential uses	
Rough service gloves and over gauntlets 	Rough service PPE gloves are designed to protect the hands from abrasion, impact, vibration, puncture, and cut hazards that are common in heavy-duty work environments. Rough service PPE gloves are usually made of durable materials that have high tensile strength, such as leather, synthetic leather, Kevlar, cotton, or nylon. Specific gloves, such as linesman or electrician's leather over gauntlets, can be used over the top of high-voltage electrically insulated gloves as added protection against abrasion or damage. The gloves provide no extra protection against electric shock, so they should be used in addition to, but not instead of, high-voltage insulated gloves.
High-voltage insulated gloves 	High-voltage PPE gloves are designed to protect the hands from electric shock when working with high-voltage systems. High-voltage PPE gloves are usually made of rubber or synthetic materials that have high dielectric strength, which means they can resist the flow of electric current. High-voltage PPE gloves are supplied in different class categories based on the maximum voltage they are rated to withstand, ranging from Class 00 (500 volts) to Class 4 (36,000 volts). ➤ The minimum class required for working on and around the high-voltage systems of electric vehicles is Class 0. ➤ High-voltage PPE gloves should be worn with leather protectors over them to prevent damage from abrasion, cuts, or punctures. ➤ High-voltage PPE gloves need to be tested before each use to ensure they are free of defects or holes that could compromise their electrical insulation. The gloves can be inflated at the wrist, and then rolled up to ensure that no air leaks are present. ➤ When a new pair of gloves are commissioned, the date they are brought into use should be recorded. Once exposed to air and UV, the materials begin to degrade. Most glove manufacturers will stipulate that the gloves will require recertification or replacement after 6 months. This can sometimes be extended to twelve months if they are not used frequently and are kept according to the manufacturers' recommendations in their original packaging. Cotton inner-liner gloves may also be used, which reduce perspiration and increase hygiene. However, they provide no extra protection against electric shock, so they should be used in addition to, but not instead of, high-voltage insulated gloves.

Electric Vehicle Hazards and Maintenance

Table 3.3 Examples of personal protective equipment PPE and potential uses	
Eye protection	Eye protection is available in various formats. It is primarily designed to provide protection against impact, heat, chemicals, and fumes. The most appropriate form of eye protection should be chosen for the type of task being undertaken. **Safety glasses** are a type of protective eyewear designed to protect your eyes from small flying objects, such as splinters or dust, as well as chemicals and light. They are designed to meet special requirements that normal glasses do not, such as: ➤ They are made of impact-resistant materials that can withstand high-speed or high-mass impacts without shattering or scratching. ➤ They have side shields or wraparound frames that cover the entire eye area and prevent objects from entering from the sides. **Safety goggles** are a type of protective eyewear designed to protect your eyes from small flying objects, such as splinters or dust, as well as chemicals and light. This reduces the risk of injury when worn. **Safety face shields** are a type of protective eyewear designed to protect your face from small flying objects, chemicals, light, and heat. They can be worn over glasses or goggles to provide extra protection. Safety face shields are made of different materials and have different features depending on the type and level of hazards involved. Some common materials used for safety face shields include: ➤ Polycarbonate: This material is impact-resistant, transparent and can block UV rays. It can also withstand high temperatures and has good optical clarity. ➤ Acetate: This material is chemical-resistant, transparent and can block UV rays. It can also resist fogging and scratching. ➤ Steel mesh: This material is durable, breathable and can protect from sparks and debris.
Hard hats and bump caps 	Hard hats and bump caps are a type of protective helmet that are designed to protect your head from injury due to falling objects, impact with other objects, debris, rain, and electric shock. Hard hats are usually made of different materials that have different properties and resistance levels against various hazards. Some common materials include: ➤ High-density polyethylene (HDPE): This material is lightweight, durable and can withstand high temperatures, UV rays, chemicals, and electricity. ➤ Acrylonitrile butadiene styrene (ABS): This material is impact-resistant, rigid and can withstand low temperatures, UV rays, chemicals, and electricity. ➤ Polycarbonate: This material is impact-resistant, transparent and can block UV rays. It can also withstand high temperatures and has good optical clarity.

Electric Vehicle Hazards and Maintenance

Table 3.3 Examples of personal protective equipment PPE and potential uses	
Safety footwear 	Safety footwear is a type of PPE that is designed to protect your feet from injuries caused by impact, penetration, heat, cold, chemicals or electricity. Safety footwear is usually made of leather, rubber or synthetic materials that have different properties and resistance levels against various hazards. Some common features of safety footwear include: ➤ Protective toe caps: These are made of steel, aluminium, composite, or plastic materials that can withstand high forces and prevent crushing or puncturing of the toe area. ➤ Penetration-resistant midsoles: These are made of steel, textile or composite materials that can prevent sharp objects from piercing through the sole of the shoe. ➤ Anti-slip soles: These are made of rubber or polyurethane materials that have special patterns or textures that can improve grip and traction on slippery or uneven surfaces. ➤ Anti-static soles: These are made of conductive materials that can dissipate static electricity and prevent electric shocks or sparks. ➤ Electrically insulated soles: These are made of non-conductive materials that can inhibit electricity and reduce the possibility of electric shock. ➤ Heat-resistant soles: These are made of rubber or polyurethane materials that can withstand high temperatures and prevent melting or burning of the sole. ➤ Waterproof or breathable uppers: These are made of leather, synthetic or textile materials that can resist water or moisture and improve comfort and hygiene. ➤ Ankle support or padding: These are made of leather, synthetic or textile materials that can provide stability and cushioning for the ankle and heel.
High-voltage overshoes 	High-voltage overshoes are a type of protective footwear designed to protect feet from electric shock when working with high-voltage systems. High-voltage overshoes are usually made of rubber or synthetic materials that have high dielectric strength, which means they can resist the flow of electric current. High-voltage overshoes are worn over regular shoes and cover the entire foot area. Similar to high-voltage insulated gloves, the overshoes have different class categories based on the maximum voltage they can withstand, ranging from Class 00 (500 volts) to Class 4 (36,000 volts).
Ear protection 	Ear protection is a type of personal protective equipment (PPE) that protects your hearing from damage caused by loud or noisy environments. Ear protection can also prevent the entry of debris and water into the ear canal. **Earplugs** are small devices that fit inside the ear canal and block out sound. They are usually made of foam, silicone, rubber, or wax. Earplugs are suitable for low to moderate noise levels and can be disposable or reusable. **Earmuffs** are devices that cover the entire ear and form a seal around it. They are usually made of plastic, metal, or leather, with cushioned pads and a headband. Earmuffs are suitable for moderate to high noise levels.

Electric Vehicle Hazards and Maintenance

Table 3.3 Examples of personal protective equipment PPE and potential uses	
Particle masks and respirators 	Particle masks and respirators are types of personal protective equipment (PPE) that cover the nose and mouth and filter out airborne particles, such as dust, smoke, pollen, bacteria, and viruses. They can help reduce the risk of respiratory infections and diseases; however, they have different designs, features, and levels of protection. **Particle masks** are usually made of non-woven fabric or paper that can trap large particles, such as dust or pollen. Particle masks can protect the wearer from droplets and splashes of fluids, but they cannot filter out small particles, such as aerosols or vapours and smoke. **Respirators** are usually made of synthetic materials that can filter out both large and small particles. They are often reusable and should be cleaned and maintained regularly. Respirators can protect the wearer from inhaling harmful substances, but they require a proper fit test to ensure a tight seal around the face.
High visibility clothing 	High visibility clothing is a type of personal protective equipment (PPE) that is designed to make the wearer more visible in low-light or dark conditions. High visibility clothing can help prevent accidents and injuries by increasing the contrast between the wearer and the background. High visibility clothing is usually made of fluorescent or reflective materials that have different properties and resistance levels against various hazards.

Safety Rescue Poles/Hooks

A high-voltage rescue pole or hook is a device that can be used to help safely rescue a person who is in contact with a live electrical conductor, equipment, or vehicle. It consists of a long, insulated pole with a hook at the end that can be used to pull the person away from the source of electric shock. The pole is made of durable materials such as resin polyester, glass fibre, or composite, and is designed to withstand voltages up to a certain level, often around 45,000 volts.

To use a high-voltage rescue pole, the rescuer should follow these steps:
1. Wear appropriate personal protective equipment (PPE) such as gloves, boots, and helmet.
2. Check the voltage rating of the rescue pole and make sure it is suitable for the situation.
3. Approach the victim from a safe distance and angle, avoiding any contact with the live conductor or equipment.
4. Extend the rescue pole, holding it behind the hand guard, and hook it around the victim's body, clothing, or belt. Do not hook it around the neck or head.
5. Pull the victim away from the source of electric shock using the rescue pole.
6. Do not touch the victim directly.
7. Place the victim on an insulated surface and check their vital signs.
8. Call for medical assistance.

Figure 3.15 A Safety Rescue Hook

Electric Vehicle Hazards and Maintenance

First Aid

The automotive industry is a high-risk environment, and no matter what precautions are taken, there is always the possibility of accidents occurring that may lead to personal injury. The following advice is not a substitute for first aid training and will only give you an overview of the action you may need to take. You should take care when you attempt to administer first aid that you do not place yourself in danger. Be very careful about what you do because the wrong action can cause more harm to the casualty.

Good first aid always involves summoning appropriate help. Many companies will have a trained first aider on site and must have a suitably stocked first aid box.

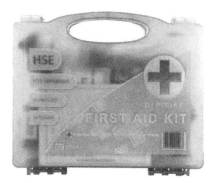

Figure 3.16 A First Aid Box

Some of the common injuries that may occur in the automotive industry are:

Cuts and lacerations: These may be caused by sharp objects, tools, or machinery.
They can range from minor to severe, depending on the depth, location, and extent of bleeding.

To treat a cut or laceration, you should:

- Apply direct pressure to the wound with a clean cloth or dressing to stop the bleeding.
- Raise the injured part above the level of the heart, if possible, to reduce blood loss.
- Cover the wound with a sterile dressing or bandage and secure it with tape.
- Seek medical attention if the wound is deep, large, or contaminated with dirt or foreign objects.

Burns: These may be caused by fire, hot liquids, chemicals, or electricity.
They can vary in severity, depending on the degree, depth, and area of the burn.

To treat a burn, you should:

- Remove any clothing or jewellery that is not stuck to the burn.
- Cool the burn with running water for at least 10 minutes or until the pain eases.
- Cover the burn with a clean cloth or dressing that does not stick to the wound.
- Seek medical attention if the burn is large, deep, or involves the face, hands, feet, or genitals.

Electric shock: This may be caused by contact with live wires, components, or faulty equipment.
It can result in burns, muscle spasms, cardiac arrest, or death.

To treat an electric shock victim, you should:

- Turn off the power source or isolate the system if possible.
- Check for breathing and pulse and start CPR if needed.
- Treat any burns as described above.
- Seek medical attention as soon as possible.

The Recovery Position

The recovery position is a way of positioning a casualty who is unconscious but breathing normally. It helps to keep their airway clear and prevent them from choking on their tongue or vomit. It also reduces the risk of further injury or complications until help arrives. However, you should not attempt to put someone in the recovery position if you suspect they have a spinal injury, or if they have a broken limb that prevents you from moving them safely.

To put a casualty in the recovery position, follow these steps:

1. Kneel beside the casualty and make sure they are lying on their back with their legs straight.
2. Place the arm nearest to you at a right angle to their body, with the palm facing up.
3. Take the other arm and place it across their chest, with the back of their hand against their cheek nearest to you.
4. Lift up the leg farthest from you and bend it at the knee.
5. Carefully roll the casualty onto their side by pulling on the bent leg. The top arm should support the head and the bottom arm should stop them from rolling too far.
6. Tilt the head back slightly to open the airway and adjust the hand under their cheek if needed.
7. Check that they can breathe easily, and nothing is blocking their mouth or nose.
8. Monitor their breathing and pulse until help arrives.
9. If they stop breathing, start CPR immediately.

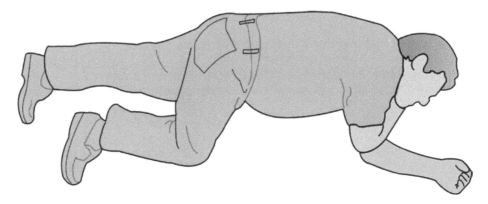

Figure 3.17 The Recovery Position

Electric Vehicle Hazards and Maintenance

 Due to the highly dangerous nature of chemicals found within batteries, any leakage of battery electrolyte must be dealt with using extreme caution and appropriate personal protective equipment (PPE) must be worn.

Chemical Spills

The **chemistry** of the **electrolyte** found in high-voltage batteries can be highly **toxic** and **caustic**. Most electrolytes are in a liquid form, meaning that if the battery is damaged or ruptured, it could cause a leak. Many electrolytes are absorbed into the other chemical **compounds** within the battery, so leaks are likely to be minor. However, any liquids released are potentially highly dangerous. For example, the electrolyte contained in a nickel metal hydride (NiMH) battery is an **aqueous** solution of potassium hydroxide. This very strong **alkaline** solution can be highly **corrosive** and is damaging to skin and eyes. If a leak is detected, special care must be taken when cleaning up any spill. Chemical resistant PPE must be worn to protect the operative and the electrolyte will need to be **neutralised** before it can be cleaned up. As potassium hydroxide is alkaline, it requires an **acid** solution to neutralise it. The neutralising agent is normally powdered boric acid, dissolved in water in the quantities recommended in any safety literature. The goal is to render the electrolyte **pH** neutral, which can be tested using **litmus paper**. The spill can then be mopped up with cloth, which should then be disposed of as **controlled waste**.

Electric Vehicle Hazards and Maintenance

 Safety product suppliers produce workshop spill kits, designed to contain the necessary equipment and chemicals to deal with a leak of electrolyte from different styles of battery. These spill kits are often supplied in a case to be used in an emergency in a similar style to a first aid kit.

Chemistry - the nature of chemical bonds in matter.

Electrolyte - a substance that can conduct electric current when dissolved in a liquid.

Toxic - containing or being poisonous material, capable of causing death or serious harm.

Caustic - able to burn or corrode organic tissue by chemical action.

Compounds - substances that are made up of two or more different types of atoms that are chemically bonded together.

Alkaline - ionic salt of an alkali metal or an alkaline earth metal, such as sodium, potassium, or calcium.

Corrosive - able to burn or corrode organic tissue by chemical action.

Neutralised - to make something ineffective by applying an opposite force or effect.

Acid - a substance that has certain chemical properties, such as reacting with metals to produce hydrogen gas, and neutralising base materials to form salts.

pH - stands for potential of hydrogen or power of hydrogen. It is a measure of how acidic or alkaline a solution is. It is a number on a scale from 0 to 14, where 7 is neutral, lower values are more acidic, and higher values are more alkaline.

Litmus paper - a type of paper that is used to test the acidity or alkalinity of a solution.

Controlled waste - a term that refers to waste that is subject to legal regulations in its handling or disposal, usually because of its potential to harm human health or the environment.

Thermal Runaway

Thermal runaway occurs when a critical breakdown of the chemistry inside a high-voltage electric vehicle battery leads to catastrophic failure. Once the chemical breakdown begins, it generates an **exothermic** reaction that is self-sustaining and **propagates** through the rest of the battery. Because of its self-sustaining nature and uncontrollable reaction, it is known as thermal runaway. Several factors can cause thermal runaway within a battery. It could be caused by overheating, **short circuit**, impact damage, or overcharging.

Exothermic - a chemical reaction or process that releases heat.

Propagate - spread and promote widely.

Electric Vehicle Hazards and Maintenance

The reaction generates large amounts of heat, often between 650°C and 800°C. A byproduct of this chemical reaction is the **thermal breakdown** and release of toxic and flammable gases, including:

- Carbon monoxide
- Carbon dioxide
- Hydrogen fluoride
- Hydrogen chloride
- Sulphur dioxide
- Hydrocarbons such as ethane, methane, butane, and hydrogen cyanide

Due to the nature of some of the gases released, this may lead to an increased risk of further fire damage and potentially explosion. Dealing with a fire caused by thermal runaway requires large quantities of water, and tackling this sort of fire may not be suitable with a standard fire extinguisher. Once the area has been evacuated, raise the alarm, and call the emergency services so that they can discharge large quantities of water across the area.

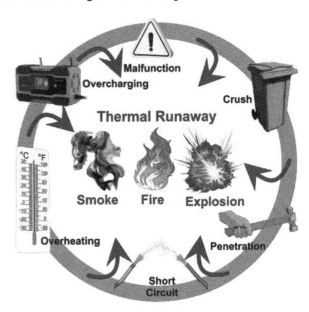

Figure 3.18 Thermal Runaway

Short circuit - a situation where an electrical current flows through an unintended path with very low resistance, causing excessive heat, damage, or fire.

Thermal breakdown - a process where a material or substance loses its integrity or functionality due to exposure to high temperatures.

Jacking Points

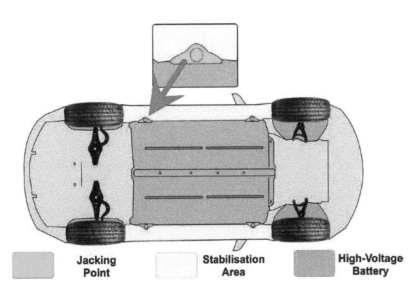

As the high-voltage battery is often mounted under the floor pan of many vehicle designs, it is very important that the manufacturers' recommendations and instructions are closely observed when raising the vehicle on a hoist or jack.

Incorrect positioning of the lifting pads could damage the battery or twist the vehicle frame, which may invalidate any warranty or safety features, and could potentially lead to a short circuit resulting in thermal runaway. Not only do manufacturers specify where a vehicle should be jacked up, but some also insist that vehicle-specific lifting pads or blocks are used.

Figure 3.19 EV Jacking Points

Electric Vehicle Hazards and Maintenance

High-Voltage Cables

The high-voltage cables found on hybrid and electric vehicles should be capable of carrying high current at high voltage with low resistance. They are often made from strands of copper or aluminium, which form the inner core, and are coated in a protective insulation, usually a form of plastic or rubber, which is designed to prevent the leakage of electrical voltage to other components or systems. The **cross-sectional area** of a cable relates to the current-carrying capability of any wiring. If the cross-sectional area of a cable is doubled, its internal resistance is halved and therefore it can carry twice as much current.

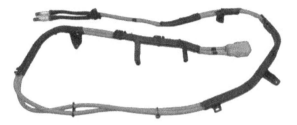

It is recommended that exposed cables, wiring, or harness connectors are colour-coded orange to indicate the potential of high-voltage within these components. Whenever orange cables, wires, or harness connectors are encountered, the operator should always assume that they are live and potentially dangerous.

Figure 3.20 High-Voltage Cables

A single strand of 0.3-millimetre-diameter copper automotive wire is capable of carrying more than 500 **milliamps**. Remember that, because 80 milliamps is often considered the fatal threshold for current, regardless of the size of the wiring or connector, if it is coloured orange, it could have a voltage potential above the touch threshold that could make current flow and therefore cause serious injury or death.

It is often suggested that the British scientist Sir Humphry Davy was the first to observe and describe the phenomenon of an electrical arc being created when a circuit is broken while current is flowing.

This phenomenon can cause injury or death if a high-voltage electric cable is disconnected from a circuit while current is flowing. It creates a situation that causes an **arc flash** or **arc blast**, which is a high-temperature form of electrical explosion. This produces vast amounts of heat, loud sounds that can damage the ears, and potentially metal spatter, where molten metal is thrown off electrical contacts.

Cross-sectional area - a term used to describe the diameter of a wire or cable when viewed end on. Increasing the cross-sectional area makes the wire thicker, while decreasing it makes the wire thinner.

Milliamps - a unit of electrical current equal to one thousandth of an ampere.

Arc flash - a dangerous phenomenon where a strong electrical current passes through the air between two conductors, releasing intense heat and light.

Arc blast - the explosive pressure wave that results from an arc fault.

Electric Vehicle Hazards and Maintenance

Inverter and Capacitors

The inverter is a component used by hybrid and electric vehicles to convert direct current (DC) from the battery to alternating current (AC) for use by the electric motors, and alternating current (AC) from the generators to direct current (DC) for recharging the battery. This is a high-voltage component and extra care should be taken when working on or around the inverter-converter unit. The inverter-converter also uses high-voltage capacitors as part of the system construction. A capacitor is a temporary storage for electrical energy, which can maintain a voltage above the touch threshold for a period of time once the vehicle has been shut down. Under normal circumstances, the power is removed from this system when the ignition is switched off and the key is removed, so routine maintenance and repairs should be safe to conduct. However, the safety systems of hybrid and electric vehicles may not be fully reliable and therefore, care must be taken when working in the vicinity of the inverter unit.

Figure 3.21 An Inverter

 It is recommended that, following the manufacturer's shutdown guidelines, a wait time should be observed before beginning any work on a hybrid or electric vehicle. This is to allow time for the capacitors in the high-voltage system to discharge through the backup or redundancy circuit.

Three-Phase Motor Generators

The drive motors and generators of hybrid and electric vehicles primarily use high-voltage alternating current, which may be boosted beyond the voltage potential of the battery. Three separate circuits, known as phases, carry individual alternating current at high-voltage to and from these units. This is why they are called 'three-phase' motor generators. Care must be taken when working on or around these areas, particularly if the motor is allowed to move or revolve, for example, by pushing or towing the vehicle. Regardless of the vehicle being switched off, movement can create electricity in the circuits, which could potentially lead to damage, injury, or death. Where possible, when undertaking maintenance and repairs of hybrid and electric vehicles, the vehicle should be stabilised to ensure that it does not move or rotate the motor generator while work is being conducted.

Figure 3.22 A Motor Generator

Vehicle Categories

Hybrid and electric vehicles can be classified into different categories [see Chapter 2] and, because of their styles and components, will require different forms of maintenance. It is important to consider these differences when planning any routine maintenance or repairs on these vehicle types.

 Never charge the HV battery of a plug-in hybrid or fully electric vehicle while maintenance or repair work is being conducted. Regardless of whether the ignition has been switched off and the key removed, the high-voltage system will be live, meaning it has the potential to injure or kill if circuits are touched.

Electric Vehicle Hazards and Maintenance

Engine Maintenance

Regardless of their design, hybrid electric vehicles require the use of an internal combustion engine as part of their propulsion system. This may be directly connected to the wheels, or act as an on-board charger for the high-voltage electrical system. However, it will require the same maintenance and servicing as with any other internal combustion engine and may involve the replacement of components such as:

- Engine oil and filters
- Spark plugs
- Air filters
- Fuel filters
- Coolant

It is important to ensure that the engine is correctly switched off and isolated before any work is carried out. The engines will operate in auto stop-start mode, meaning that when idle, the engine may appear to be switched off but could start if a form of electrical or mechanical demand is called for. Unexpected start-up may lead to injury of the technician or damage to the vehicle.

To help ensure that a hybrid or electric vehicle is switched off and isolated before any routine maintenance or servicing is conducted, the vehicle should be switched on first to ensure that no malfunction indicator lights (MIL), or warning symbols appear on the driver's instrument display.
The ignition should then be switched off and, if the vehicle uses a smart key that does not need to be inserted into a slot, this should be placed beyond its range of operation according to the manufacturer's prescribed instructions.
To ensure that the key has been placed beyond its range of operation, or that a spare key is not located elsewhere within the vehicle, the operator should try to restart the vehicle by putting it back into ready mode.
Normally, the driver display will indicate that the key is not detected and therefore give reassurance that the vehicle has shut down and is out of ready mode.

Regenerative Braking

Most hybrid and electric vehicles make use of the principle of regenerative braking to assist with the slowing down of the vehicle, and also recover some electrical energy that can be used to help recharge the high-voltage battery [see Chapter 2]. However, friction brakes are still required to make up any difference in deceleration and stopping required by the driver. To enable the hybrid or electric drive system to deal with this effectively, many manufacturers incorporate a method of brake-by-wire.

Brake-By-Wire

The regenerative braking used in hybrid and electric vehicles means that only some of the energy needed to slow the vehicle down comes from the hydraulic system. To ensure that the deceleration of the vehicle is accurately controlled, many manufacturers incorporate brake-by-wire systems in their vehicle design. Instead of the master cylinder applying hydraulic pressure directly to the brake calliper system, it operates as a pressure measurement sensor (also known as a stroke simulator.) This sensor simulates pedal pressure so that brake operation feels normal to the driver. The signal from the brake pressure sensor is processed by an ECU, which operates a secondary master cylinder, often powered by a pressure accumulator unit. In conjunction with the wheel speed sensors, the secondary master cylinder applies the appropriate brake force required to slow the wheels without allowing them to skid. If the regenerative braking and hydraulic brake-by-wire fail, the system enters conventional brake operation, which will allow the vehicle to be slowed down, but the driver may have to push the pedal slightly harder and overall stopping distances may be increased.

Electric Vehicle Hazards and Maintenance

Braking Maintenance

As with any traditional braking system, the hydraulic and mechanical components wear over a period of time and therefore require periodic maintenance. Due to regenerative braking, friction materials have a tendency to last longer than those found on traditional systems. However, the corrosion and seizing of mechanical components is often increased. This means that when conducting the service and maintenance of braking systems, particular care should be taken when assessing the free movement of all mechanical and hydraulic components.

Brake friction material, i.e. pads and discs, will need to be assessed for wear, contamination, or damage when conducting servicing and replaced as necessary, following the manufacturer's recommendations.

Hydraulic brake fluid deteriorates over time, and mineral-based fluids that are hygroscopic will require periodic replacement due to the absorption of water moisture, which will lower its boiling point and promote corrosion of metallic hydraulic components.

As with other forms of maintenance on hybrid or electric vehicles, the ignition should be first isolated, and the vehicle shut down. However, the bleeding of the hydraulic system will often require that the vehicle is powered up and placed in a maintenance mode. To ensure the correct operation and safety, always follow the manufacturer's procedures.

Tyres

Vehicle tyres have five main functions:

> Transmit driving and braking forces - Driving forces from an engine or electric motor need to be transmitted to the road surface through the contact of the tyre and the road surface. When slowing down, the braking effort must be efficiently transmitted in the opposite direction, often contributing to regenerative braking.

> Assist with steering - The lateral force exerted on a tyre's sidewall when the steering is turned will help a vehicle when cornering or manoeuvring.

> Support the load of the vehicle - The tyre casing and the pneumatic nature of the air inside must be able to adequately support the weight and load of the vehicle.

> Provide traction and grip in all weather conditions - The tread patterns manufactured into the surface of the tyre provide channels designed to squeeze water out of the way and ensure the tyre surface remains in contact with the road. This helps reduce the possibility of aquaplaning.

> Provide a smooth ride for the vehicle's occupants - The pneumatic construction of the tyre can act like a form of spring, absorbing some of the road shocks created by uneven surfaces.

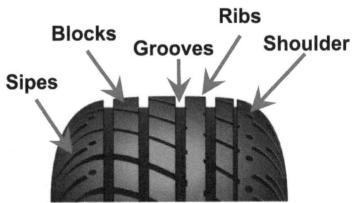

Figure 3.23 Tyre Tread

Electric Vehicle Hazards and Maintenance

A tyre is required to produce friction and grip. By its very nature, this creates a condition known as rolling resistance. Rolling resistance will have a direct effect on the economy and efficiency of hybrid and electric vehicles. Some manufacturers use tyres with a different design to ensure that the contact patch between the tyre and the road surface remains similar, but rolling resistance is reduced. This is achieved by making the tyre taller and narrower.

The contact patch between the tyre tread and the road surface on many vehicles is approximately the size of an A4 piece of paper in landscape. If this contact patch is rotated into a portrait direction by making the tyre taller and narrower, the area of contact can remain the same, but this will often provide a reduced rolling resistance.

The legal requirement for minimum tyre tread depth will vary depending on vehicle type and geographical location. However, during any maintenance, the tread depth and condition of the tyre should always be assessed to ensure safety and conformity with any legal requirements.

Tyre pressures have a direct impact on safety, but also on rolling resistance and therefore on efficiency and economy. Tyre pressures should always be maintained at the manufacturer's specific recommendations.

Cooling Systems

Regardless of whether the vehicle is hybrid or fully electric, cooling systems are required in order to maintain the correct operation and effectiveness of engines or fully electric drive systems. A coolant, often combined with some form of antifreeze and corrosion inhibitors, will be used for this purpose. A hybrid vehicle may have a separate cooling system for its internal combustion engine and electric drive motor system. An electric vehicle will have a cooling system to maintain the temperature of electrics and electronics found in the inverter and motor generator systems. During routine service and maintenance, this coolant should be checked for level, condition, and effectiveness. Coolant may require periodic replacement depending on the manufacturer's instructions.

Figure 3.24 Vehicle Cooling Systems

Electric Vehicle Hazards and Maintenance

Some manufacturers may use a waterless coolant. This will be a synthetic blend of chemicals specifically designed to support the type of system used and should not be topped up or mixed with any water-based coolant.
Waterless cooling systems can sometimes be identified, as the reservoir is often sealed to prevent accidental contamination with a water-based antifreeze. Always use the manufacturer's specific coolants designed for the system being protected.

Cooling systems are often pressurised to raise the boiling point of the liquid coolant. A cooling system should never be opened while hot and pressurised. The sudden reduction in pressure can cause the coolant to boil and be released as superheated steam, causing injury.

Some manufacturers provide an insulated reservoir to maintain coolant at a higher temperature when the vehicle has been switched off. This means that the warm-up time of an internal combustion engine, for example, can be reduced. Extra care must be taken when accessing these systems, as stored temperature could also lead to boiling of coolant and injury to the operator.

Heating Ventilation and Air Conditioning HVAC

The maintenance of the heating, ventilation, and air conditioning (HVAC) systems in electric vehicles is often seen as a simple task, but it's actually related to the high-voltage system. Regardless of whether the vehicle uses an internal combustion engine or electric drive, any maintenance should only be conducted by a suitably qualified F-gas technician following all legal and environmental rules.

Specialist training and certification is required to legally service and maintain these systems.

Electric vehicles and some hybrids need a way to heat the passenger compartment because they might not have engine-heated coolant and a heater matrix. Instead, they use a positive temperature coefficient (PTC) thermal resistor. This electric heating element warms air or liquid, which is then used to heat the cabin.

Cabin heating elements in electric vehicles and hybrids are powered by the high-voltage drive battery. This can drain the battery and affect the vehicle's range. To address this, some manufacturers offer alternatives like electrically heated seats and steering wheels. These draw power from the low-voltage auxiliary system, reducing the load on the high-voltage drive battery, while ensuring passenger comfort.

Air conditioning and climate control systems in vehicles help regulate temperature, ventilation, humidity, and air purity. These systems are primarily based on refrigeration processes and may be connected to other systems to provide all necessary features.

Electric Vehicle Hazards and Maintenance

In hybrid or electric vehicles, the air conditioning and climate control work similarly to conventional systems, with one key difference. The compressor in a conventional system is powered by the internal combustion engine, which can't operate with electric drive or during stop/start on hybrid motors. Instead, the compressor is powered by an electric motor from the high-voltage battery.

Because the compressor uses the high-voltage system, it requires a special non-electrically conductive lubricating PAG (polyalkylene glycol) or POE (polyolester oil). When servicing the air conditioning systems of an electric or hybrid vehicle, it is important to use only the recommended oil.

Air Conditioning Components

In order to function, the air conditioning system is made up of a number of components.

Table 3.4 illustrates the main parts used and gives a brief description of their purpose.

Table 3.4 The main components of an air conditioning system	
Air conditioning component and function	**Example**
Compressor: This is a high-voltage electric pump designed to increase the pressure of the refrigerant in the system.	
Condenser: This is a type of radiator located outside the passenger compartment, typically in front of the engine radiator. Its function is to cool the hot refrigerant gas and allow it to condense into a liquid.	
Evaporator: This is a type of radiator located inside the passenger compartment, typically just in front of the heater matrix. Its function is to allow the refrigerant to expand and evaporate, changing state from a liquid to a gas.	
Receiver Drier: This is a container designed to serve as a reservoir for liquid refrigerant. It contains a silicone desiccant, which is hygroscopic, to help remove any water moisture present in the system. Receiver driers are used in Thermal Expansion Valve (TXV) air conditioning systems.	
Expansion Valve: Also known as a Thermal Expansion Valve or TXV, this is a variable-sized nozzle that controls the amount of refrigerant entering the evaporator. It regulates the flow based on the evaporator's temperature, allowing more refrigerant in as the temperature rises and reducing the flow as the temperature falls.	

Electric Vehicle Hazards and Maintenance

Table 3.4 The main components of an air conditioning system	
Suction Accumulator: This is a container that temporarily stores liquid refrigerant. It allows the refrigerant to expand and evaporate before it returns to the compressor. It contains a silicone desiccant, which is hygroscopic, to help remove any water moisture present in the system. Suction accumulators are used in Fixed Orifice Tube (FOT) air conditioning systems.	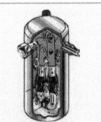
Fixed Orifice Tube (FOT): This is a precisely sized restriction located at the entrance to the system evaporator. It ensures a constant, metered flow of refrigerant during air conditioning operation.	
Hoses: These are rubber connectors that link various parts of the system. Rubber is used to minimise damage from vibrations between components mounted solidly to the vehicle body and those mounted on the engine or drive motor assembly. To ensure proper operation, these hoses are made from rubber compounds that are resistant to chemical damage from refrigerants and help prevent refrigerant leakage.	

The systems operate based on the principle of the refrigeration cycle. This cycle is a process of heat transfer. It works by absorbing heat energy from the vehicle's passenger compartment as air circulates through the evaporator and transferring it to the external atmosphere as it passes through the condenser radiator located outside the vehicle. The heat energy isn't destroyed; it's simply moved from inside the vehicle to the outside.

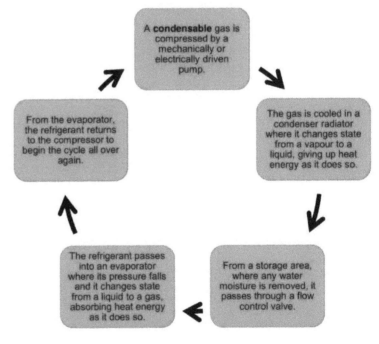

Figure 3.25 The Refrigeration Cycle

Condensable - a substance that can change state from a gas or vapor to a liquid.

Electric Vehicle Hazards and Maintenance

Expansion Valve Type System (TXV)

The TXV air conditioning system uses a high-voltage electric pump, known as a compressor, to increase the pressure of a refrigerant gas within a sealed system. The gas then travels through a radiator, referred to as a condenser, typically positioned just in front of the cooling system radiator. Here, some of the heat generated by compression is removed, slightly cooling the high-pressure gas, and condensing it into a liquid. This liquid is then transferred to a storage container called a receiver drier until needed.

When the driver adjusts the controls to lower the temperature in the vehicle's passenger compartment, the refrigerant is released through a temperature-controlled expansion valve (TXV). As the pressure drops, the liquid refrigerant changes state in another small radiator inside the vehicle, known as the evaporator. The temperature in the evaporator decreases, and as the cabin air circulates through the evaporator's fins, heat is removed. This process cools the air inside the vehicle. The refrigerant then returns to the compressor, and the entire process starts all over again.

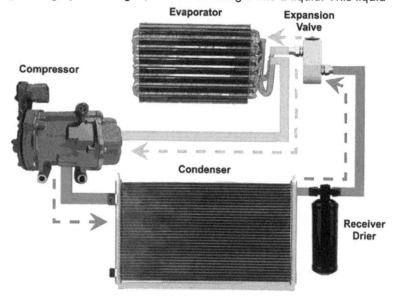

Figure 3.26 A TXV System

Fixed Orifice System (FOT)

An alternative air conditioning system to the Thermal Expansion Valve (TXV) type uses a process with a Fixed Orifice Tube (FOT).

In this system, a high-voltage electric pump, known as a compressor, is used to increase the pressure of a refrigerant gas within a sealed system. The gas then travels through a radiator, referred to as a condenser, typically positioned just in front of the cooling system radiator. Here, some of the heat generated by compression is removed, slightly cooling the high-pressure gas, and condensing it into a liquid.

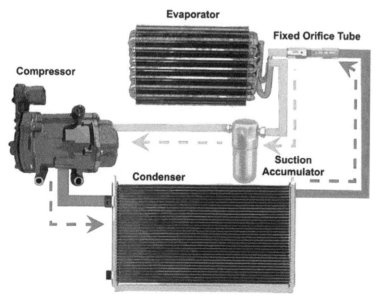

After the condenser, the gas passes through an accurately sized restriction called a 'fixed orifice' into the evaporator. In the evaporator, pressure drops allowing the refrigerant to change state back into a gas, and the temperature decreases. As the cabin air circulates through the evaporator's fins, heat is removed, which cools the air inside the vehicle.

When the cold vapour leaves the evaporator, it enters a temporary reservoir called a suction accumulator. There, any liquid refrigerant that may have passed through the evaporator is separated from the vapour before being returned to the compressor. The vapour then returns to the compressor, and the entire process starts all over again.

Figure 3.27 A FOT System

Electric Vehicle Hazards and Maintenance

Some manufacturers also use the heat transfer properties of the air conditioning refrigeration cycle to directly support the thermal management of the high-voltage batteries.

If the batteries are air-cooled, with the input air source located inside the passenger compartment, air conditioning indirectly affects them by drying the air and reducing corrosion on battery connections and components.

Safety Precautions and Procedures When Working on EV/Hybrid Vehicles

Working on or around the high-voltage systems of hybrid and electric vehicles is potentially dangerous, even though routine maintenance or servicing carries the same risks as a traditional vehicle.

Some additional hazards that should be considered are listed in **Table 3.5**.

Table 3.5 Hazards associated with high-voltage electrical systems found on vehicles	
Hazard	**Risk or danger**
Electric shock	Electric shock is caused when electricity passes through the human body, leading to injury or death. The risk of electric shock increases as electrical voltage rises. According to the **ECE R100** regulations, voltages higher than 60 volts DC or 30 volts AC **RMS** values increase the risks posed. However, the voltage touch threshold may be lower than those stated in this regulation.
Burns	A common effect of electric shock on the human body is burning. This is due to the fact that the electrical energy is converted to heat as it is discharged. Burns caused by electrical discharge through the body may be internal and cause tissue damage that might not be immediately apparent. External burning to the skin can also occur if an accidental short circuit is created during the connection and disconnection of a high-voltage electrical circuit, as the heat given off can be intense.
Arc flash	Arc flash is caused by a sudden, accidental discharge of electrical energy that creates an arc similar to that of welding. The temperatures created when this occurs can be around 20,000 degrees Celsius, which is enough to vaporise clothing and human flesh. Arc flash is most likely to occur at voltages above 480 volts, which are not uncommon on electric vehicle drive systems.
Arc blast	Arc blast is an explosion caused by a short circuit in a high-voltage system. It has many of the same effects as arc flash, such as causing severe burns, but also has the potential to cause injury at a distance. Added dangers may involve flying particles from the system and a sound pressure wave that might lead to permanent hearing damage.
Fire	A rapid discharge of electrical energy will create large amounts of heat. This heat will then have a high potential of igniting flammable materials. When working on high-voltage systems, it is advisable to have a suitable fire extinguisher to hand (one that will not conduct electricity).
Explosion	High-voltage systems often combine electricity with chemicals, gases and fumes that pose a risk of explosion. Care should always be taken to remove any sources of ignition and correctly isolate electrics before working on these systems.

Electric Vehicle Hazards and Maintenance

ECE R100 - a regulation that defines safety requirements for electric powertrains and electric vehicles.

RMS - stands for root mean square and it is a way of measuring the effective value of an alternating current (AC) or voltage. It roughly equates to the peak-to-peak average of an AC wave and is proportional to a DC voltage equivalent.

Responsibilities of Management and Skilled Employees

Most responsibilities stem from Health and Safety at Work legislation, but all activities involving electric and hybrid vehicles should be risk assessed in advance.

Employer's Responsibilities

It is an employer's duty to protect the health, safety and welfare of their employees and other people who might be affected by their business.

Employers must do whatever is reasonably practicable to achieve this.
This means making sure that workers and others are protected from anything that may cause harm, effectively controlling any risks to injury or health that could arise in the workplace.

Employers have duties under health and safety law to assess risks in the workplace. Risk assessments should be carried out that address all risks that might cause harm in a workplace.

Employers must provide information about the risks in your workplace and how you are protected, and also instruct and train you on how to deal with the risks.

Employers must consult employees on health and safety issues. Consultation must be either direct or through a safety representative that is either elected by the workforce or appointed by a trade union.

Employee's Responsibilities

All workers are entitled to work in environments where risks to their health and safety are properly controlled. Under health and safety law, the primary responsibility for this lies with employers.

Workers have a duty to take care of their own health and safety and that of others who may be affected by their actions at work.

Workers must co-operate with employers and co-workers to help everyone meet their legal requirements.

Legislation is the process of making or enacting laws.

- Specific and non-specific legislation will always need to be observed and followed; however, the legislation will vary depending on your geographical location.
- It is important that you are aware of the legislation and your rights and responsibilities, as well as those of your employer.
- It is your right to expect your employer to fulfil their responsibilities and it is your employer's right to expect you to fulfil yours.

Legislation is the law and, if you do not observe it, you are committing an offence.

Electric Vehicle Hazards and Maintenance

External Charging

[Information regarding electric vehicle charging modes can be found in Chapter 2].

Charging an electric or plug-in hybrid vehicle while it is in the workshop undertaking service or maintenance presents several hazards.

> - If a vehicle is on charge, regardless of whether it has been switched off and the key removed beyond its range of operation, the high-voltage system must be live and awake in order for the charging to take place. This means that because the vehicle cannot be shut down, no service or maintenance should be undertaken while the vehicle is on charge, regardless of how minor the activity is considered.
>
> - Secondly, trailing leads or charging cables not only present a trip hazard, but also theoretically could provide an alternative path to ground and therefore a circuit that increases the potential risk of electrocution.
>
> - Extension leads should not be used, as they may not be rated for the current flow required during charging and if left wound up as used in storage, current flow and strong magnetic fields could lead to overheating, melting and fire. Extension leads or any trailing leads should not be allowed to get wet.

Figure 3.28 Electric Vehicle Charging

If work is to be conducted, for example service or maintenance, the vehicle should be disconnected from charge.

If the vehicle is on charge in a workshop, the key should be removed, and signs and barriers placed so that everybody knows not to touch or work on the vehicle while it is on charge.

Summary

This chapter has described:

- Safety and systems of work.
- Hazard management.
- Legislation and regulations.
- Types of fire extinguishers and suppressants.
- Personal Protective Equipment (PPE) and first aid.
- Electric vehicle hazards and maintenance.

High-Voltage Component Replacement

Chapter 4 High-Voltage Component Replacement

In this chapter, you will learn about the high-voltage components of electric vehicles (EVs). You will be provided with an overview of the high-voltage systems that power EVs, as well as the procedures and best practices for replacing high-voltage components. The chapter will also cover the safety measures and precautions that must be followed when working with high-voltage systems, such as wearing personal protective equipment (PPE), using insulated tools, and following the shutdown, isolation, and lockout procedures.

Electric vehicles are powered by batteries that store electrical energy and deliver it to the drive motors. The batteries operate at high-voltages, typically working with several hundred volts, which require special care and precautions when handling them. High-voltage components, such as the battery pack, inverter, motor generators, and cables, are essential for the performance and safety of EVs. However, they may also degrade over time or malfunction due to various factors, such as temperature, humidity, vibration, corrosion, or physical damage. Therefore, it is important to know how to inspect and replace high-voltage components in electric vehicles.

By reading this chapter, you will gain a better understanding of the potential hazards and actions that could be implemented as good working practice. You will also learn how to interact with these vehicles safely and efficiently, whether you are a driver, a passenger, a technician, a member of the automotive industry, or a first responder. This chapter is important for those who are replacing electric vehicle high-voltage components.

Contents

- ❖ Information sources **Page 108**
- ❖ Safety precautions and procedures when working on EV/Hybrid vehicles **Page 108**
- ❖ Battery technology **Page 117**
- ❖ Removing/refitting and testing high-voltage components **Page 130**
- ❖ High-voltage components **Page 132**
- ❖ Electric vehicle charging systems **Page 147**

The automotive industry is a high-risk environment, especially when dealing with electrical systems. The hazards of electricity are well-known, but they can be easily ignored due to its invisible nature. This can lead to complacency unless the fundamental operation of electric vehicles is understood. Even with this understanding, caution is necessary. Do not rely on any safety systems designed for protection; instead, take precautions to minimise the risk of injury or death. Always evaluate the risks associated with any activity and implement measures to eliminate or reduce the hazards involved in any task that involves hybrid or electric vehicles.

Additional risks associated with working on, or around electric vehicles may include:

- ➤ Electrocution
- ➤ Strong magnetic fields
- ➤ Falling from height
- ➤ Injuries caused by incorrect manual handling techniques
- ➤ Chemicals
- ➤ Gases or fumes
- ➤ Hybrid engine systems starting unexpectedly
- ➤ The silent movement of electric vehicles while in use

Never attempt to work on a high-voltage electrical system unless you have received adequate training.

High-Voltage Component Replacement

Information Sources

The replacement of high-voltage components on hybrid and electric vehicles requires a reliable source of technical information and data. Before starting and during the repair process, it's crucial to have comprehensive information about these systems.

Potential sources of information may include:

Table 4.1 Possible information sources	
Verbal information from the driver	Vehicle identification numbers
Service and repair history	Warranty information
Vehicle handbook	Technical data manuals
Workshop manuals/Wiring diagrams	Safety recall sheets
Manufacturer specific information or data sheets	Information bulletins
Technical helplines	Advice from other technicians/colleagues
Internet	Parts suppliers/catalogues
Jobcards	Diagnostic trouble codes
Oscilloscope waveforms	On vehicle warning labels/stickers
On vehicle displays	Temperature readings

Always compare the outcomes of any high-voltage component replacement against appropriate data sources. Regardless of the information or data source you choose, it's crucial to assess its usefulness and reliability for your repair work.

If you need to replace any electrical or electronic components, always ensure that the quality meets the original equipment manufacturer (OEM) specifications. If the vehicle is under warranty, using inferior parts or making deliberate modifications might invalidate the warranty. Additionally, fitting parts of inferior quality could affect vehicle performance and safety. You should only replace electrical components if the parts comply with the legal requirements for road use.

Safety Precautions and Procedures When Working on EV/Hybrid Vehicles

Working on or around the high-voltage systems of hybrid and electric vehicles is potentially dangerous. Hazards include, but are not limited to:

- Electric shock
- Burns
- Arc flash
- Arc blast
- Fire
- Explosion

[For details of these hazards, see **Table 3.5**, Chapter 3]

High-Voltage Component Replacement

Safety Precautions to be Taken Before Carrying Out Any Repair Procedures on Electric Vehicles

The typical voltages used for a range of electrically propelled and hybrid vehicles are 100 to 800V.

Table 4.2 gives examples of recommended high-voltage PPE that should be used when working with electric vehicle drive systems.

Table 4.2 Examples of high-voltage personal protective equipment PPE	
PPE	**Recommendations and use**
High-voltage insulated gloves and over gauntlets	High-voltage PPE gloves are designed to protect the hands from electric shock when working with high-voltage systems. High-voltage PPE gloves are usually made of rubber or synthetic materials that have high dielectric strength, which means they can resist the flow of electric current. ➤ High-voltage PPE gloves are supplied in different class categories based on the maximum voltage they are rated to withstand, ranging from Class 00 (500 volts) to Class 4 (36,000 volts). The minimum class required for working on and around the high-voltage systems of electric vehicles is Class 0 (1,000 volts). ➤ High-voltage PPE gloves should be worn with leather protectors over them to prevent damage from abrasion, cuts, or punctures. However, they provide no extra protection against electric shock, so they should be used in addition to, but not instead of, high-voltage insulated gloves. ➤ High-voltage PPE gloves need to be tested before each use to ensure they are free of defects or holes that could compromise their insulation. The gloves can be inflated at the wrist, and then rolled up to ensure that no air leaks are present. ➤ When a new pair of gloves is commissioned, the date they are brought into use should be recorded. Once exposed to air and UV, the materials begin to degrade. Most glove manufacturers will stipulate that the gloves will require recertification or replacement after 6 months. This can sometimes be extended to twelve months if they are not used frequently and are kept according to the manufacturers' recommendations in their original packaging. ➤ Cotton inner-liner gloves may also be used, which reduce perspiration and increase hygiene. However, they provide no extra protection against electric shock, so they should be used in addition to, but not instead of, high-voltage insulated gloves.
Eye protection	Eye protection is available in various formats. It is primarily designed to provide protection against impact, heat, chemicals, and fumes. The most appropriate form of eye protection should be chosen for the type of task being undertaken. ➤ Safety glasses are a type of protective eyewear designed to protect your eyes from small flying objects, such as splinters or dust, as well as chemicals and light. They usually have side shields or wrap around the temples to prevent objects from entering from the sides. ➤ Safety goggles are a type of protective eyewear designed to protect your eyes from small flying objects, such as splinters or dust, as well as chemicals and light. They fit tightly against the face and form a seal around the eyes, preventing any particles or liquids from entering. ➤ Safety face shields are a type of protective eyewear designed to protect your face from small flying objects, chemicals, light, and heat. They can be worn over glasses or goggles to provide extra protection.

High-Voltage Component Replacement

Table 4.2 Examples of high-voltage personal protective equipment PPE	
Overalls or Workwear 	Specific workwear or overalls provide an additional layer of protection between the user and potential hazards. Depending on the materials used, this protection could include containment of loose clothing and the covering of areas of bare skin, resistance to chemicals or limited fire protection. It is recommended that overalls or clothing do not use metal fastenings when working on or around the high-voltage systems of hybrid and electric vehicles. Metal fastenings may increase the risk of short circuit, leading to electrocution or fire, as well as interfere with the very strong magnetic fields found within the drive systems of electric vehicles.
	Arc flash/blast overalls are a type of PPE that is designed to protect the wearer from the hazards of arc flash and blast, which are intense bursts of heat, light and sound that can occur when an electrical fault causes an arc between two conductors. Arc flash and blast can cause severe burns, blindness, hearing loss, shock and even death. Arc flash/blast overalls are usually made of flame-resistant fabrics that do not ignite, melt, or drip when exposed to high temperatures. They may also have reflective strips or patches to increase visibility in low-light conditions.
Safety footwear and high-voltage overshoes 	Safety footwear is a type of PPE that is designed to protect your feet from injuries caused by impact, penetration, heat, cold, chemicals or electricity. Safety footwear is usually made of leather, rubber or synthetic materials that have different properties and resistance levels against various hazards. Some common features of safety footwear include: ➤ Protective toe caps: These are made of steel, aluminium, composite, or plastic materials that can withstand high forces and prevent crushing or puncturing of the toe area. ➤ Penetration-resistant midsoles: These are made of steel, textile or composite materials that can prevent sharp objects from piercing through the sole of the shoe. ➤ Electrically insulated soles: These are made of non-conductive materials that can inhibit electricity and reduce the possibility of electric shock. They are also often marked with a green triangle symbol to indicate their electrical safety rating.
	High-voltage overshoes are a type of protective footwear designed to protect feet from electric shock when working with high-voltage systems. High-voltage overshoes are usually made of rubber or synthetic materials that have high dielectric strength, which means they can resist the flow of electric current. High-voltage overshoes are worn over regular shoes and cover the entire foot area. Similar to high-voltage insulated gloves, the overshoes have different class categories based on the maximum voltage they can withstand, ranging from Class 00 (500 volts) to Class 4 (36,000 volts). The minimum class required for working on and around the high-voltage systems of electric vehicles is Class 0 (1,000 volts).

High-Voltage Component Replacement

Other safety precautions that should be observed when working with vehicle high-voltage systems are shown in **Table 4.3**.

Table 4.3 Safety precautions when working on high-voltage systems	
Hazard	**Precautions**
Using electrical test equipment with high-voltage systems	Wear appropriate high-voltage PPE. Where possible, only use electrical hand tools that are specifically designed for use with high-voltage systems. Ensure test equipment is correctly rated for the high-voltage system you will be working on - normally a minimum of CAT III 1000 volts. Conduct a visual inspection of the meter and test leads/probes to ensure condition and that there is no apparent damage. When using a multimeter, ensure that your fingers are behind the insulating finger guards of the test probes. Calibrate voltmeters by testing them on a known good voltage, such as the 12 volt battery or a proving unit, to ensure they are working correctly and accurately. If the multimeter screen shows a low battery indicator, replace the battery immediately. A low battery in the multimeter may lead to incorrect readings being taken, which could result in injury or death.
Disposal of waste materials	Under the Environmental Protection Act (EPA), you must treat old batteries (lead acid, lithium ion and metal hydride) as hazardous waste and dispose of them in the correct manner. They should be safely stored in a clearly marked container until they are collected by a licensed recycling company. This company should give you a waste transfer note as proof of collection. High-voltage components should only be disposed of by returning them to an authorised recycler. Often, the vehicle or original equipment manufacturer (OEM) is able to accept high-voltage component returns or will recommend an approved disposal method.
Dealing with leakage	Wear appropriate chemical-resistant PPE. If battery leakage occurs, cover the spill with a neutralising agent, for example, dilute boric acid for use with alkaline electrolytes. After neutralising, rinse the contaminated area with clean water. If the spill involves a large amount of electrolyte, call the fire services, and allow them to handle it. This may help prevent you from getting seriously hurt and reduce any environmental issues. Safety product manufacturers are able to supply spill kits which can be used in the event of an accidental leakage of battery electrolyte.

High-Voltage Component Replacement

Table 4.3 Safety precautions when working on high-voltage systems	
High-voltage electrical system isolation	When working on or around the vehicle's high-voltage components, you must correctly isolate and insulate the system. This should involve: ➤ Making others aware that the high-voltage system is being worked on. ➤ Cordoning off the area with barriers and high-voltage warning signs. ➤ Switching off and removing the ignition key (if the ignition key cannot be removed, due to damage for example, take out all of the fuses in the fuse boxes). ➤ If the ignition key is a 'smart key' which doesn't require insertion into a key slot, it should be safely stored beyond its recommended range of operation. ➤ Disconnecting the low voltage (12v) auxiliary battery if accessible. ➤ Checking and wearing appropriate high-voltage PPE. ➤ Following manufacturer's instructions, removing the high-voltage service plug, switch, or high-voltage interlock loop (HVIL) and 'locking-out' so that they cannot be accidentally reconnected. ➤ Allowing the specified time for any system capacitors to discharge and checking the voltage has fallen to a safe level using an appropriately calibrated and rated voltmeter. ➤ Not cutting any orange high-voltage cable. ➤ Isolating any disconnected terminals with insulating high-voltage coverings.
Submerged vehicle safety	To handle a hybrid or electric vehicle that has been partially or fully submerged in water, the high-voltage system and airbags will need to be safely isolated. Wear fully insulating electrical PPE [as described in **Table 4.2**]. Then immobilise the vehicle and remove the ignition key. (If the key is a 'smart key', ensure it is kept beyond its recommended range of operation.) Next, allow any water to drain or dry if possible. Finally, remove all system fuses and isolate the high-voltage system as described above.
Highly magnetic components	The magnets used in hybrid and electric vehicle drive motors are around 10 to 15 times stronger than standard iron magnets. The naturally high magnetic field produced can affect the correct operation of life-sustaining electronic medical equipment. Care should be taken to remove all metal jewellery and avoid the use of delicate electronic equipment, such as mobile phones, while working on these systems. If the rotor of a hybrid or electric drive motor needs to be removed or fitted, it will be necessary to use special tooling, so that the magnetic attraction to other metal components does not affect its fitment and create damage to the motor or cause personal injury. Existing medical conditions, such as heart conditions, can be affected by both the very strong magnetic fields produced in hybrid and electric drive motors and other high-voltage systems. It is not recommended that people with electronic medical life-sustaining equipment, such as heart pacemakers, automatic electronic defibrillators (AED) or insulin pumps, for example, work on these systems.

High-Voltage Component Replacement

Table 4.3 Safety precautions when working on high-voltage systems	
Checking voltage prior to working near or on high-voltage systems	Before any work is started on a high-voltage system, the electrical circuits must be isolated, and capacitors allowed to discharge. The system voltage should be checked with a correctly rated and calibrated multimeter to ensure that it has fallen low enough to begin any work. Voltages higher than 60 volts DC or 30 volts AC RMS are likely to cause electric shock leading to injury or even death. However, even though the risk of electrocution below these values has diminished, the risk of short circuit remains. Therefore, a safe working voltage should be considered as close as possible to zero.

High-Voltage Tools and Equipment

Table 4.4 High-voltage tools, equipment, and safety	
Tool/Equipment	**Description and use**
Signs and Barriers	An important step of any high-voltage electric vehicle maintenance or repair is ensuring that others are aware of the work that is being conducted. High-voltage warning signs should be placed on and around the vehicle, and safety barriers or chains should be erected to create a buffer zone which will keep others more than arm's length away from the vehicle being worked on. Additional safety posters should be displayed in the workshop describing emergency or first aid procedures relating to high-voltage.
Fire Extinguishers	Although most extinguishers will be unsuitable for use on a high-voltage battery electric fire, the workshop must have fire extinguishers available designed to tackle different types of fire [see Chapter 3]. Specialist extinguishers such as aqueous vermiculite dispersants (AVD) are available to help tackle fires involving lithium ion, and other suppressants can also help contain chemical fires. It is important that if a fire should occur, the alarm is raised, the immediate area is evacuated, and the emergency services are called.
Fire Blanket	Some safety suppliers have produced fire blankets specifically for use with electric vehicle fires. Although these blankets are not designed to extinguish the fire, they can be used to cover and smother the entire vehicle, as long as it is safe to do so, helping to contain the fire and reduce damage. These blankets may provide time to effectively evacuate the area and limit collateral damage until the emergency services arrive.

High-Voltage Component Replacement

Table 4.4 High-voltage tools, equipment, and safety	
Lockout Kits 	Lockout kits provide covers, storage boxes and padlocks that are designed to secure the vehicle and high-voltage components, reducing the possibility of someone accidentally touching components they should not, or reconnecting any high-voltage system while work is being conducted. Lockouts are designed to be visible components, with warning labels stating the high-voltage danger. Often made of plastic, they can be used to cover steering wheels, ignition switches, isolator plug sockets and high-voltage interlock loops. Key boxes are available that are designed for storing ignition keys beyond the recommended range of operation.
Faraday Key Pouch	As many electric vehicles use a smart key that does not require placing in a slot or ignition keyhole, a Faraday pouch is an additional component that can be used to help shield radio signals from the key. Named after the scientist Michael Faraday, this pouch is specifically designed to help store keys and prevent the transmission of a wireless signal that may allow the vehicle to be started or placed in ready mode accidentally.
Safety Rescue Hook 	A high-voltage rescue hook is a vital part of electric vehicle safety equipment. Often with a composite or fibreglass handle insulated to around 45,000 volts, a hook is formed on one or both ends to allow the rescuer to drag a victim away from a source of electrocution. If this is not possible, the hook could also be used to drag high-voltage components or wiring away from the victim or isolate the high-voltage source. The rescuer must be wearing high-voltage personal protective equipment (PPE) at all times, and not place themselves in any danger.
First Aid Kit 	Good first aid always involves summoning appropriate help; many companies will have a trained first aider on site and must have a suitably stocked first aid box. The minimum level of first aid equipment in a suitably stocked first aid box should include: ➤ A guidance leaflet. ➤ 2 sterile eye pads. ➤ 6 triangular bandages. ➤ 6 safety pins. ➤ 3 extra-large, 2 large and 6 medium-sized sterile unmedicated wound dressings. ➤ 20 sterile adhesive dressings. It is important to ensure that the contents of the first aid box are in date and are sufficient, based on the assessment of the workplace's first aid needs. The law does not state how often the contents of a first aid box should be replaced, but most items, especially sterile ones, are marked with expiry dates.

High-Voltage Component Replacement

Table 4.4 High-voltage tools, equipment, and safety	
Eye Wash Station 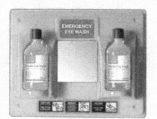	An emergency eye wash station provides sterile liquid that can supply a gentle flow of water to rinse the eyes in case of exposure to harmful substances or materials. It is critical safety equipment that can prevent or reduce eye damage and vision loss. An emergency eye wash station should be located near the potential hazards and be easily accessible. It should also be regularly inspected and maintained to ensure proper functioning.
Automatic Defibrillator AED 	An automatic defibrillator is a type of Automated External Defibrillator (AED) that checks the heart rhythm and can send an electric shock to the heart to restore a normal rhythm during sudden cardiac arrest. The operation and use of an automatic defibrillator are normally defined by instructions attached to or displayed on the unit itself. Automatic defibrillators are primarily designed for use by untrained or inexperienced bystanders, as they do not require any decision making or manual action from the user. They are suitable for environments where anyone can access and use them in an emergency.
Electrolyte Spill Kit 	The electrolytes used in the high-voltage drive battery of vehicles contain chemicals that are highly toxic and caustic, meaning that if they come into contact with human skin, poisoning or severe chemical burns may occur. Although very rare, a spill of electrolyte will require the use of an emergency spill response kit. Safety product suppliers produce kits that are suitable to deal with the electrolytes found in lead acid, lithium, and nickel metal hydride batteries. They will contain a high-performance acid or alkaline neutraliser, absorbent pads, test strips to ensure that the acid or alkali has been neutralised, and disposal sacks. They will often also contain comprehensive instructions and manufacturers' specific data sheets. If dealing with the chemical spill of electrolyte, the operator must always be wearing fully chemical resistant personal protective equipment (PPE).
Insulated Component Covers and Sleeves/Shrouds 	Exposed high-voltage components, cables and terminals should be covered with insulation that is specifically designed for this purpose while the vehicle is being worked on. This way, the possibility of accidental contact with high-voltage parts that could lead to the risk of electrocution is reduced. Often brightly coloured, or with warning labels printed on their surface, these insulators give an indication to others that the potential voltage of these components is dangerous, and they should not be touched.

High-Voltage Component Replacement

Table 4.4 High-voltage tools, equipment, and safety	
Insulated Floor Mat 	An additional layer of protection against electrocution that can be used while working on high-voltage electric vehicles is the use of insulated floor mats. These rubberised mats are specifically designed and rated to ensure that they provide insulation against high-voltage electricity. These are often rated as class zero 1000 volts. They can be placed on the floor, and the technician or operator can stand on them while work is being conducted. Insulated floor mats should conform to specific standards, for example IEC 61111. They should be kept clean, free from dirt, grease, or moisture and stored correctly when not in use to prevent any potential damage.
High-Voltage Insulated Hand Tools 	When working on or around the high-voltage electrical components, connectors, cables, and terminals, insulated high-voltage electric hand tools should be used. These tools are designed to meet specific standards and provide insulation protection up to around 1000 volts. They should conform to standards and regulations such as IEC 60900. The primary function of the insulation on the hand tools is to reduce the possibility of short circuits when working on or around high-voltage systems, although it may provide some protection to the user. Specific high-voltage personal protective equipment (PPE) must always be worn as well.
CAT III Multimeter 	In order to assess the absence of voltage when replacing components on electric drive systems, it is important to have a good quality, correctly rated and calibrated multimeter. A minimum of category three (CAT III) 1000 volts is required to assess the electrical systems of high-voltage vehicles. This meter should be checked before use for condition, including the leads and probes, and calibrated on a known good voltage source such as the 12 volt auxiliary battery or a proving unit.
Insulation Tester 	The high-voltage system of an electric vehicle is fully insulated from the vehicle's frame or chassis. When you replace high-voltage components, check them to ensure that no breach of insulation is created that may lead to a leakage of electrical voltage to the vehicle's frame. Use an insulation tester for this purpose. It is a form of resistance tester, designed to measure high levels of electrical resistance. Often containing capacitors, these can be charged to various high-voltage levels and connected between the high-voltage system and the vehicle chassis, and a resistance test can be conducted. An extremely high resistance, often in the Mega or Giga ohms, indicates sufficient insulation. Always consult the manufacturers' instructions for voltage levels and resistance results.

High-Voltage Component Replacement

Battery Technology

Low Voltage Lead Acid

A standard lead-acid battery, the type often used for vehicle auxiliary systems, consists of a number of sections called cells [see **Figure 4.1**]. Each cell can produce approximately 2.1 volts. A standard battery has six cells connected in series, creating a battery with a voltage potential of 12.6 volts. This is often rounded down, so we say that the battery has a voltage of 12 volts.

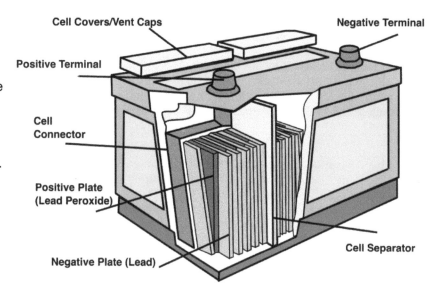

Figure 4.1 A Lead Acid Battery

Each cell contains several lead plates, which have different chemical compositions:

> The negative plate is made of lead.
> The positive plate is made of lead dioxide.

To prevent the plates from touching each other and causing a short circuit, thin sheets of material called separators are inserted between them. An **electrolyte**, which consists of sulphuric acid and deionised water, fills the space between the plates within the cell and allows **ions** to move between the plates. When connected in a circuit, the electrons flow from the negative plate to the positive plate. This creates an electric current and provides the energy to power components in the vehicle. The top of each plate is then connected to the rest of the circuit. The circuit must contain a load to consume the electrical potential energy. When the circuit is complete, the electrons react with the electrolyte and move from one plate to another as a chemical reaction, generating current.

 The Latin name for lead is plumbum (Pb) and this is how it is represented on the periodic table of elements. This name is also the origin of the word plumbing.

The chemical formula of lead dioxide is PbO_2. It is also known as lead peroxide. It is an inorganic compound that is a dark-brown solid and is insoluble in water.

Why Batteries Need to be Recharged

Batteries work with electricity flowing in one direction only (direct current). This means that eventually both plates in a low voltage lead acid battery will undergo a chemical reaction and become the same substance (lead sulphate) and the current will stop.
When this happens, no more electricity flows through the circuit, meaning that the battery is discharged. When the battery is discharged, it needs to be recharged.

In a conventional vehicle, an alternator (which can be thought of as a generator for electricity) is driven by the engine.
To recharge the battery, an electric voltage is supplied to the battery circuit with a magnitude that is higher than the electromotive force (**EMF**) (approximately 12.6 volts with all electrical consumers switched off). This reverses the chemical reaction – it pushes electrons back through the electrolyte to their original positions in the lead and lead dioxide plates and recharges the battery.

High-Voltage Component Replacement

In a hybrid or electric vehicle, the high-voltage batteries will also need recharging. In these systems, the electric drive motor can act as a generator when it is reversed.

A hybrid or electric vehicle may not have an alternator to recharge the low-voltage battery and support the auxiliary electrics, so an alternative system is required. This system consists of a drive motor/generator that will first charge the high-voltage batteries that power the electric drive system. The voltage is then stepped down through a DC-to-DC converter, where it is applied to the low-voltage battery circuit.

Electrolyte - an electrically conductive material that is usually a liquid or a gel. Electrolytes contain ions that can move freely and carry electric charges.

EMF - electromotive force, also known as voltage, which is the difference in electric potential between two points. EMF is often measured when no current is flowing in a circuit.

Ions - molecules that have a net positive or negative electric charge. They are formed by losing or gaining electrons, or by breaking or joining bonds with other particles.

Table 4.5 explains some terms related to lead-acid batteries and some of their types and ratings.

Table 4.5 Battery terms and ratings	
Term/Rating	**Description**
Ampere-hour (Ah)	A measurement of the electrical current that a battery can deliver. This quantity is one indicator of the total amount of charge that a battery can store and deliver at its rated voltage. The ampere-hour (Ah) value is the product of the discharge current (in amperes) and the time (in hours) for which this discharge current can be sustained by the battery. For example, a vehicle battery rated as 100 Ah should contain enough electricity to provide: ➤ 100 amps for one hour ➤ 1 amp for 100 hours ➤ 10 amps for 10 hours Any other combination that multiplies together to make 100 (e.g. 25 amps for 4 hours). The ampere-hour rating is required by law in Europe to be shown on a battery.
Cranking Amps (CA)	A number that represents the amount of current a lead-acid battery can deliver at 0°C (32°F) for 30 seconds and maintain at least 1.2 volts per cell (7.2 volts for a 12-volt battery).
Cold Cranking Amps (CCA)	A number that represents the amount of current a lead-acid battery can provide at −18°C (0°F) for 30 seconds and maintain at least 1.2 volts per cell (7.2 volts for a 12-volt battery). This test is more demanding than those conducted at higher temperatures.
Hot Cranking Amps (HCA)	A number that represents the amount of current a lead-acid battery can provide at 27°C (80°F) for 30 seconds and maintain at least 1.2 volts per cell (7.2 volts for a 12-volt battery).
Reserve Capacity Minutes (RCM), also known as Reserve Capacity (RC)	A lead-acid battery's ability to sustain a minimum stated electrical load, which can continuously deliver 25 amperes before its voltage drops below 10.5 volts. It is defined as the time (in minutes) that the battery at 27°C (80°F) can deliver this performance.

High-Voltage Component Replacement

Absorbed Glass Mat (AGM) Low-Voltage Battery

Some manufacturers use a form of lead acid battery to power the low-voltage vehicle systems, which is known as an absorbed glass mat (AGM) battery. In this type, the electrolyte is held in a glass fibre mesh, which acts as separators between the plates. By containing the electrolyte in the glass mat, the amount of hydrogen gas given off during charging is considerably reduced. The battery is vented to external air to reduce any excessive gas build-up during operation and ensure that any internal pressures are kept within acceptable limits. Because the electrolyte of an AGM battery is not in a free liquid state, it cannot be topped up, so the battery is sealed for life in a maintenance-free (MF) design. AGM batteries have higher power density, lower self-discharge rate, and longer lifespan than conventional lead acid batteries.

 AGM batteries need careful charging because of their construction. Excess voltage and current can damage the internal components easily. Use an approved battery charger for external charging.

Low-Voltage Earth Return Systems

Vehicle designers and manufacturers try to keep the amount of wiring used to a minimum. This will save on materials, improve efficiency, and reduce costs. Because many vehicles are manufactured mainly from metals (which are good conductors of electricity), it is not always necessary to complete an electrical circuit back to the low-voltage battery using wire alone.

> You can connect the negative end of a low-voltage electrical circuit wiring to the vehicle body or chassis. This is called an earthing point.
> You can also connect the negative terminal of the low-voltage battery to the vehicle body or chassis to complete the circuit. This is called **earth return**.

Disconnecting and Connecting the Low-Voltage Battery

When you connect or disconnect the low-voltage battery from a vehicle, you should remove and refit the terminals in a certain order, as this will help reduce the possibility of a short circuit.

> Disconnecting order: Always remove the negative lead first when you are disconnecting the battery. Once you have disconnected the negative terminal, the vehicle's electrical system is now **open circuit**. Therefore, if the tool you are using accidentally touches the vehicle's bodywork, they have the same electrical potential or pressure. If the pressure in an electric circuit is equal, no current can flow (because a difference in pressure creates flow).
> Connecting order: When reconnecting a low-voltage battery, always connect the positive terminal first and the negative terminal last, for the same reasons.

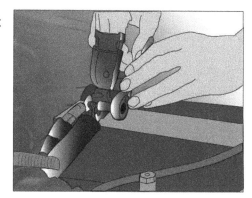

Figure 4.2 Disconnecting a Low-Voltage Auxiliary Battery

 **Open circuit** - a broken electrical circuit where no electricity can flow.

Earth return - using the metal chassis or frame to complete the vehicle's electrical circuit.

High-Voltage Component Replacement

Connecting or disconnecting electrical components on hybrid and electric vehicles can cause extreme damage to the vehicle and personal injury. If components or systems are left switched on, excessive current draw can generate heat. Always ensure that no power is discharged through an electric circuit or component during the connection or disconnection of an additional 12 volt power source.

A small amount of **parasitic drain** can cause the low-voltage auxiliary system battery to drop below its regulated operating value due to its small capacity. This would prevent the electric vehicle from entering a ready state, similar to having a flat battery on a traditional internal combustion engine vehicle. You can jump-start the low-voltage auxiliary system in a similar way to a traditional internal combustion engine vehicle, but you need to observe several precautions. The low-voltage auxiliary battery is often hidden from view and might be secured behind interior panelling, which may require some dismantling. Some manufacturers put the low-voltage auxiliary battery inside the main casing with the high-voltage battery, which makes it hard to access. They also have sophisticated battery monitoring electronics to control any charge or discharge, which are not common in the auxiliary system of traditional internal combustion engine vehicles. Therefore, manufacturers often provide dedicated jump-starting points in the engine or motor compartment area and indicate them in the driver's operating manual. Connect jump leads or a jump pack to these jump points using the correct **polarity**. Connect positive to positive and negative to negative. Do not connect polarity incorrectly as this can damage the electric vehicle's auxiliary system irreversibly. After connecting the jump leads or jump pack, put the vehicle in ready mode and disconnect the jump leads or jump pack carefully to avoid arcing or short circuit. Unlike jump-starting a traditional petrol or diesel vehicle, which needs large amounts of electric current to crank a starter motor, the low-voltage auxiliary load is taken up by the DC-DC converter once a hybrid or electric vehicle is in ready mode. A voltage potential high enough to close the high-voltage contactors is all you need to power the vehicle up.

Two forms of external jump starter are often available.

> - A jump pack contains a heavy-duty battery that can supply a large amperage at the required voltage for the auxiliary system.
> - A boost pack is smaller and contains a lightweight battery that charges a large **capacitor**. The stored energy can then be quickly released into the auxiliary system to start the vehicle. On-board electronics prevent accidental discharge of the capacitor that can cause electric shock or damage when incorrectly connected to a circuit. They only operate when a system load is detected. If a boost pack is used to jump-start a hybrid or electric vehicle and the jump points provide the circuit for the DC-DC converter, the system may not sense circuit load and power the vehicle if it is powered down. In this case, you can use a boost pack directly on the terminals of the 12 V auxiliary system, but make sure to connect with correct polarity and avoid damaging any sensitive battery monitoring electronics.

High-Voltage Component Replacement

Parasitic drain - a term used in the automotive world to describe a situation where the vehicle's battery continues to discharge even when the ignition is switched off.

Polarity - the state of having positive and negative electric charges or magnetic poles.

Capacitor - an electronic device that stores electrical energy in an electric field by accumulating electric charges on two closely spaced surfaces that are insulated from each other.

Nickel Cadmium Battery

Nickel cadmium (NiCad) batteries are rarely used for road vehicles, but they may be found in some electric vehicles. They work in a similar way to standard lead-acid batteries, but they require less maintenance and cannot be overcharged. This type of battery is made of the following materials:

- Positive plates: nickel hydroxide
- Negative plates: cadmium
- Electrolyte: potassium hydroxide and water

These batteries tend to be larger and more expensive than normal lead-acid batteries. However, they are better at coping with the extreme loads placed on them by some electrical systems, especially in utility vehicles.

Nickel-Metal Hydride Battery (Ni-MH)

Nickel-metal hydride (NiMH) batteries used to be the most popular for hybrid drive vehicles' high-voltage systems. Nickel-metal hydride batteries are similar to nickel-cadmium batteries but use a metal hydride (hydrogen atoms stored in metal) as the negative plate. A NiMH battery can have two or three times the capacity of an equivalent nickel-cadmium battery, meaning smaller size but more electrical storage. A single cell of a nickel-metal hydride battery produces 1.2 V, and they connect in series, often in groups of six to form packs called modules or blades. The modules then connect in series to form the complete high-voltage battery. Current hybrid drive systems (depending on manufacturer) use batteries with a high-voltage potential between 100 V and 300 V. A nickel-metal hydride battery has these advantages over other battery types:

- They have a high electrolyte conductivity, which allows them to be used in high power applications (such as hybrid drive).
- The battery system can be sealed, which minimises maintenance and leakage issues.
- They operate over a very wide temperature range.
- They have very long-life characteristics when compared with other battery types – this offsets their higher initial cost.
- They have a higher energy density and lower cost per watt than other battery types.

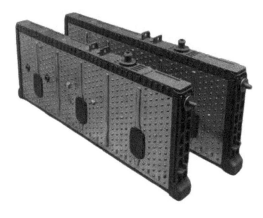
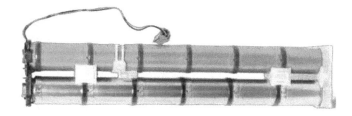

Figure 4.3 Nickel Metal Hydride Battery Modules

High-Voltage Component Replacement

Because of the characteristics of a Nickel-Metal hydride battery, the charging and discharging have to be very carefully monitored and controlled. If the battery pack is allowed to charge too quickly, overheating and damage can occur. The hybrid drive generator system will endeavour to maintain a constant current to **trickle charge** the high-voltage battery to try and sustain a set **state of charge (SoC)**. An ideal state of charge is around 60% of battery capacity, allowing the battery to work well within its capabilities.

Battery temperature is carefully monitored because as the cells become fully charged any excess energy will be converted into heat. **Thermistors** are used to measure battery temperature and if required an electronically controlled fan is able to draw air through ducting surrounding the high-voltage battery unit and assist with cooling.

Trickle charge - a slow charging method that is equal to or slightly above the battery's natural discharge rate.

State of charge (SoC) - a rating that shows how much electricity is contained in the battery compared to its capacity.

Thermistors - temperature-sensitive resistors.

The measurement of temperature in the high-voltage battery system can also indicate the battery's state of charge (SoC). By monitoring temperature, the battery system ECU can send signals to the generator regulation control unit and adjust the amount of charge supplied.

Lithium Ion Battery

For fully electric vehicles, a battery type other than Nickel-Metal hydride that can deliver, and store high-voltage electricity is needed. Lithium ion (Li-ion) batteries have a higher energy density, (the amount of energy they hold by volume or by weight) than many other battery types. Generally, the Li-ion battery cells hold roughly four times as much energy as the Nickel-Metal hydride (NiMH) batteries that are used in some hybrids. Their internal chemistry makes them work in a slightly different way than other battery types. Lithium is an extremely reactive element with only one electron in its outer shell (orbit). When brought into contact with many other types of elements, it immediately wants to give away its outer electron and become a positively charged ion. Keeping a balance between the lithium in a stable state and in a reactive state is key to the operation of the lithium ion battery.

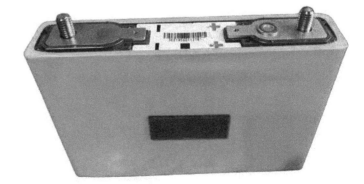

Figure 4.4 Lithium Ion Cells

High-Voltage Component Replacement

A cell is created where the lithium is in a stable state as part of a metal oxide. The cell also contains a layer of graphite that can be used as a storage medium for the unstable lithium and an electrolyte that only allows lithium ions to pass through.

When an external potential (charge) is applied to the metal oxide, the lithium ions are attracted towards the positive side of the cell, while the outer shell electrons from the lithium pass through the external circuit from the cell and then rejoin the lithium in the graphite storage area. At this point the lithium cell is charged with a voltage potential of approximately 3.6 volts (depending on the design and additional chemicals used by the battery manufacturer).

When connected to a circuit containing an electrical load, the lithium migrates back to its stable state in the metal oxide. The lithium ions can pass through the electrolyte; however, the free electrons pass through the external circuit and load, with the resulting current flow, powering the attached consumer. [See Chapter 5].

Lithium Ion battery cells are very susceptible to over and under charge. They need to be kept within a certain voltage range; otherwise, irreversible damage occurs. Voltage needs to be monitored and maintained between 2.4 and 4.2 volts per cell using sophisticated battery management.

Operational Cell Voltage - Energy – Temperatures

Table 4.6 provides some nominal voltages, energy density, temperature values, and charging advice for different battery types.

Table 4.6 Battery operational information					
Battery type	Nominal voltage (per-cell)	Approximate energy mass/volume	Charge temperature	Discharge temperature	Charging advice
Lead acid	2.1 V	35 Watt/hours per Kilogram 80 Watt/hours per litre	−20°C to 50°C (−4°F to 122°F)	−20°C to 50°C (−4°F to 122°F)	Charge at 0.3C-rate or less below freezing. Lower voltage threshold by 3mv/°C when hot. Charge at 0.1C-rate between −18°C and 0°C.
NiMH	1.2 V	125 Watt/hours per Kilogram 350 Watt/hours per litre	0°C to 45°C (32°F to 113°F)	−20°C to 65°C (−4°F to 149°F)	Charge at 0.3C-rate between 0°C and 5°C. Charge acceptance at 45°C is 70%. Charge acceptance at 60°C is 45%.
Li-ion	3.6 - 3.75 V	200 Watt/hours per Kilogram 250 Watt/hours per litre	0°C to 45°C (32°F to 113°F)	−20°C to 60°C (−4°F to 140°F)	No charge permitted below freezing. Good charge/discharge performance at higher temperature but shorter life.

The C-rate is the unit used to measure the speed at which a battery is fully charged or discharged. For example, charging at a C-rate of 1C means that the battery is charged from 0% to 100% in one hour.

High-Voltage Component Replacement

Series and Parallel Construction

To form a battery pack, individual cells are grouped together and connected to each other in a **series** or **parallel** format.

> When connected in series, the voltage of each individual cell is added together to provide a total battery pack voltage.

> When connected in parallel, the capacity of each individual cell is added to the total battery pack capacity (the amount of energy the total battery can hold).

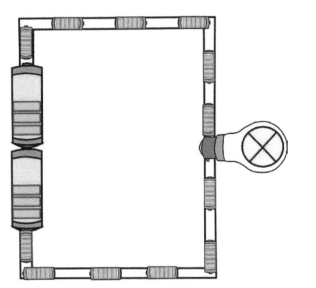

Figure 4.5 Battery Cells Connected in Series

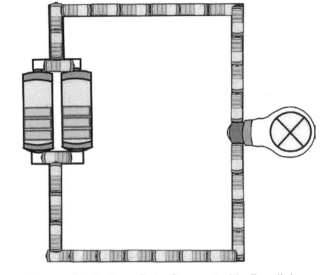

Figure 4.6 Battery Cells Connected in Parallel

Series - connected one after another.

Parallel - connected side-by-side.

When documenting or providing information about a battery pack or module, series construction is often referred to by using the letter S and parallel construction is often referred to using the letter P.

For example: 6S - 6P would mean that a pack or module contains six series-connected cells and six parallel-connected cells.

Cooling/Thermal Management

A byproduct of the charging and discharging process that occurs within a high-voltage battery while in use is the generation of heat. Thermal management of some kind is required to keep the battery within its safe operating limits. As cells become fully charged, the generation of heat can lead to gassing, damage, or thermal runaway. Thermistors are temperature sensors that are strategically placed within the battery construction to monitor heat and, if necessary, engage some form of cooling.

High-Voltage Component Replacement

Thermal Management

Thermal management is often achieved using one of the following three methods:

Air circulation - An electrically powered fan controlled by the battery management unit can draw air through the battery casing via ducting and assist with battery cooling.

Figure 4.7 An Air-Cooling System

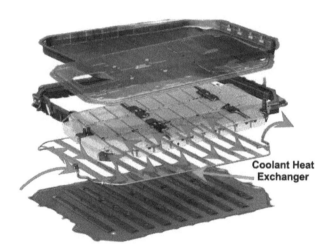

Liquid cooling - A dedicated liquid cooling system can circulate coolant through a heat exchanger inside the battery casing. An electrically powered coolant pump controls the speed and flow of coolant and thus the amount of cooling provided.

Figure 4.8 A Liquid-Cooling System

Air conditioning - a subsystem of the vehicle's air conditioning can be used with its own dedicated evaporator, to extract heat from inside the battery casing while circulating cold air around the battery cells.

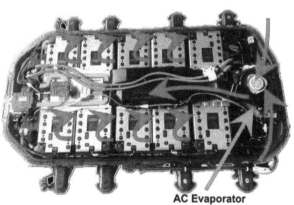

Figure 4.9 An AC Cooling System

With an air-cooling system, air is often drawn from inside the vehicle cabin to reduce the humidity level, helping to prevent corrosion on electrical battery terminals and bus bars. The vehicle's internal air conditioning system is an indirect method of moisture removal.

Vehicles designed for use in extremely cold environments may have battery heaters to reduce the possibility of cell electrolyte freezing. Positive-temperature coefficient (PTC) heating pads, often driven from the high-voltage battery itself, will raise the temperature of the cells by a very small amount. For example, if the temperature drops to minus 17 degrees Celsius, the electrical heating pads could be used to raise it to minus 10 degrees Celsius to prevent the extreme cold.

High-Voltage Component Replacement

Chemical Risks

Due to the highly toxic and caustic nature of the chemical electrolytes found in high-voltage batteries, it is important to handle any potential leaks appropriately [see Chapter 3]. Safety suppliers produce spill kits that can be used in the event of an electrolyte leak, while wearing correct chemical-resistant personal protective equipment (PPE). Chemical spills should be risk assessed, and any waste materials should be disposed of following environmental procedures and policies.

Figure 4.10 An Electrolyte Spill Kit

Fire and Explosion Risks - Thermal Runaway

Battery cells containing an **aqueous electrolyte** will convert into gases if overcharged. One of the main gases produced is hydrogen, which is highly flammable. If exposed to a source of ignition, this hydrogen gas may catch fire and, if contained in a confined space, may lead to a rapid pressure expansion and possible explosion. Therefore, charging and discharging must be carefully controlled, and some manufacturers include venting in their battery designs to direct hydrogen away from the battery and the vehicle.

Lithium-ion batteries are highly **volatile** and therefore at risk of overheating, causing an **exothermic** reaction which generates heat leading to fire and/or explosion. This reaction is known as **thermal runaway** or **thermal explosion**. The main causes of thermal runaway include:

- Overcharging
- Rapid discharge
- Short circuit
- Overheating
- Poor manufacture
- Damaged cells due to accident

Thermal runaway creates extreme heat ranging from 600 to 900 degrees Celsius. It also produces gases including hydrogen fluoride, which is flammable and highly toxic.

Misusing a battery can cause it to overheat, explode, or ignite, resulting in serious injury. It's crucial to avoid exposing the battery to fire or heat, installing it with reversed polarity, connecting the terminals with a metal object, or subjecting it to strong impacts or shocks. You should also avoid placing the battery near high-temperature locations.

These actions could cause the battery to generate heat, explode, or ignite.

Where possible, firefighting should be left to the professionals. Trying to tackle a fire with a standard fire extinguisher can be highly dangerous and possibly make the situation worse. It is necessary to discharge a large amount of water to cool the battery and area, but if left exposed to oxygen, residual heat could cause the fire to reignite. [See Chapter 3 for more information about battery fires and firefighting].

Aqueous - made from, with, or by water.

Electrolyte - a substance that can conduct electricity when it is dissolved in water or melted.

Volatile - something that can change quickly, unpredictably, or violently.

Exothermic - a chemical reaction that releases heat to the surroundings.

High-Voltage Component Replacement

Thermal runaway - a process that happens when a system produces more heat than it can dissipate, leading to a rapid and uncontrollable increase in temperature.

Thermal explosion - a type of chemical reaction that occurs when a system produces more heat than it can dissipate, leading to a rapid and uncontrollable increase in temperature. This can cause dangerous or destructive outcomes, such as explosions, fires, or melting.

When working on or near the high-voltage system of an electric vehicle, it is possible to isolate the battery from the rest of the vehicle. However, the battery itself can never be shut down. Due to its chemical nature, the battery will always remain live and, although certain precautions can be taken, there is always the risk of short circuit, electrocution, injury, or death.

Proving Units

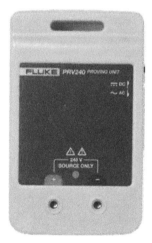

A correctly calibrated and rated voltmeter is an essential tool for checking the high-voltage system and proving absence of voltage before beginning any work. It is important to ensure that the voltmeter and test leads are at least Category 3 (CAT III) 1000 volts. The multimeter and leads should be inspected to ensure they are undamaged and in good condition. The voltmeter should be tested on a known good voltage source to calibrate and ensure correct operation before use, such as the 12-volt auxiliary battery.

A proving unit is another good method for conducting this test. A proving unit is a portable device that often contains capacitors that can be charged to a preset value. When the probes of the voltmeter are connected across the test connections of the proving unit, the reading can be checked against the specified value.

Figure 4.11 A Proving Unit

A good test method will involve the 'live-dead-live' procedure. To conduct this:

- ☑ Check the voltmeter using the proving unit (live).
- ☑ Check the high-voltage circuit for absence of voltage (dead).
- ☑ Check the voltmeter again on the proving unit to ensure that everything is working correctly and that reliable measurement readings have been obtained (live).

Proving units are available in several different formats. They can be either alternating current (AC) only, direct current (DC) only, or switchable between AC and DC. It is important to ensure that the correct proving unit, current, and values are used for the circuit to be tested.

High-Voltage Component Replacement

Systematic High-Voltage Battery Isolation Approach

In order to shut down or isolate the high-voltage system of a hybrid or electric vehicle, manufacturers' procedures should always be adhered to. However, the following approach can be used as a systematic procedure to help ensure safety.

Before disconnecting the low-voltage auxiliary battery, conduct a voltage test with a voltmeter; this will confirm several things:

> The voltmeter is working, accurate, and set to the correct measurement unit.
> The vehicle has shut down and switched off the DC-to-DC converter. (If charging voltage is seen, the high-voltage system is still awake. Do not disconnect the low voltage auxiliary battery.)
> The operator can assess the State of Charge (SoC) of the low-voltage auxiliary battery.

1 • To ensure everybody knows that work is being conducted, signs and barriers should be placed around the vehicle, creating a buffer zone to help keep people away. This buffer zone should extend beyond arm's length of others.

2 • The vehicle should be powered up and placed in ready mode to ensure that no malfunction indicator lights are present, which may show a fault with the vehicle. The vehicle should then be switched-off, key removed and placed beyond its range of operation if it is a smart key.

3 • If accessible, the negative terminal of the 12 volt auxiliary battery should be disconnected and isolated so that it cannot accidentally reconnect. [See diagnostic tip.]

4 • Before any high-voltage isolators or components are touched, high-voltage PPE including Class 0 gloves and a face shield should be inspected, tested and worn. [See **Table 4.2**].

5 • Following manufacturers' instructions, remove the high-voltage isolator, maintenance service disconnect (MSD) and place it in a secure location so that it cannot accidentally be reconnected. If the vehicle has a high-voltage interlock loop (HVIL), this should be disabled and secured using a lockout to ensure that it cannot accidentally be reconnected.

6 • The operator should then wait the recommended time to allow capacitors to discharge.

7 • Following manufacturers' instructions, the calibrated multimeter should then be used to check for absence of voltage from the high-voltage battery and capacitors.

Figure 4.12 Isolating a High-Voltage Electric Vehicle

High-Voltage Component Replacement

An additional precaution that can be used during the shutdown and isolation procedure is to try and place the vehicle in ready mode again after the smart key has been placed beyond its range of operation. This will normally show 'Key not detected' on the driver's display. This helps to prove that:

> The vehicle is switched off.
> The key is beyond its range of operation.
> There is no spare key located inside the vehicle.

Insulated Tools and Their Importance in Relation to Short Circuit

It is essential to use correctly rated and insulated hand tools when working on or around the high-voltage systems of an electric vehicle. The classification of these tools is normally conducted by the International Electrotechnical Commission (IEC), which is an organisation that publishes the standards for electrical and electronic related technologies. These tools should be classified and tested to IEC 60900, which is applicable to insulated, insulating and hybrid hand tools used for working live or close to live parts at nominal voltages up to 1000 V AC and 1500 V DC. These tools can often be identified by a 1000 V (double triangle) IEC EN 60900 mark with the name of the manufacturer, the tool reference, and the year of manufacture.

Some insulated tools may be classified under the German VDE standard Verband der Elektrotechnik, Elektronik und Informationstechnik (Association for Electrical, Electronic and Information Technologies).

The main purpose of insulated tools is to reduce the possibility of electrical short circuit when working on or around the high-energy electrical systems. A short circuit may cause arcing, rapid heating, fire, or even explosion, leading to injury or death.
The insulation performs a secondary function, which can be an additional protection to the user or operator against electrocution. However, insulated tools should always be used in addition to high-voltage electrical personal protective equipment (PPE) and never instead of.

Figure 4.13 Insulated Hand Tools

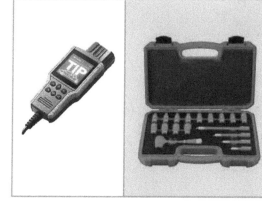
Although the colour coding of insulated hand tools is not standardised, they are often coated in red or orange coverings to indicate their potential use with high-voltage systems.

Some manufacturers provide a double coating for their tooling, first in yellow and then covered with red insulation. This means that the user or operator can easily identify any damaged or worn-out tooling. If the tool wears or is damaged, the yellow under insulation is exposed.

High-Voltage Component Replacement

Removing/Refitting and Testing High-Voltage Components

Once the high-voltage battery has been successfully **isolated**, the capacitors have been discharged and the absence of voltage has been proven, components can be replaced.

When replacing any high-voltage electrical or electronic components, always ensure that the quality meets the original equipment manufacturer (OEM) specifications. Fitting second-hand or inferior parts may affect vehicle performance, reliability, and safety. You should only replace high-voltage electrical components if the parts are supplied from a reputable source and comply with the legal requirements for road use.

Although the high-voltage battery has been isolated, it is recommended that insulated tools be used on any high-voltage electrical components and connections to reduce the possibility of accidental short circuit causing damage, injury, or death.

You must always adhere to the manufacturer's specifications, including specific **torque** settings for mounting bolts and especially any electrical connections. Most high-voltage electrical connections have relatively low torque settings and damage, or poor connection may occur if these are overlooked.

Before reconnecting a high-voltage component and powering up the vehicle, you should conduct an insulation test using an insulation tester (megohmmeter) to ensure that no breach of insulation has been created to the vehicle frame or chassis while work has been undertaken. [See Chapter 5 for a description of insulation testing].

Tests to be Conducted When Repairing or Replacing High-Voltage Electric Components

Following any repairs or replacement of high-voltage electrical components, diagnoses and tests, you need to conduct checks to ensure the serviceability, safety and operation of the electrical systems.

Indirect tests using vehicle information systems and/or live data should be perfectly safe for the operator to conduct. However, you must always view the reliability and accuracy of any data obtained on its merits and the source of information provided.

Only people who have had sufficient training and experience should conduct physical measurements using diagnostic test tools, especially if 'live' high-voltage readings or checks are required.

Diagnostic equipment and electrical test measurement tools must be correctly rated for the voltages to be tested, calibrated, and checked against a known good voltage source for accuracy.

You must always wear high-voltage personal protective equipment (PPE).

Isolated - a deliberate non-conductive separation between two parts of an electrical circuit.

Torque - a measurement of turning effort.

High-Voltage Component Replacement

Table 4.7 describes some potential tests that could be conducted following the repair or replacement of high-voltage electrical components.

[See Chapter 5 for details of test methods and procedures].

Table 4.7 Potential tests to be conducted following the repair or replacement of high-voltage components	
Test/Check	Overview
Visual	You should conduct a visual inspection of any high-voltage components to be repaired or replaced to ensure integrity, conformity, suitability, and safety. You should assess components for damage, contamination, wear and tear. Any warning, guidance or labels should be intact and legible.
Voltage	A correctly rated and calibrated voltmeter can be used to directly check components and supply values. For testing high-voltage vehicle systems, the multimeter must be a minimum of CAT III 1000 volts with matching probes and leads, tested against a known good voltage source. Live high-voltage tests should only be conducted by sufficiently trained and experienced operators wearing appropriate personal protective equipment (PPE).
Ready Mode	When switching on, vehicles perform a self-test before allowing the vehicle to enter ready mode. As long as safety systems have not been compromised, and issues or faults are electrically detectable, powering up the vehicle and ensuring it is able to enter ready mode can be used as a low-level indication of correct component function and operation.
Malfunction Indicator Lights (MIL)	Most vehicle driver information systems include warning indicators to alert the operator of any current or pending faults or issues. Malfunction indicator lamps (MIL) should initially illuminate as a self-check and ensure that they function, and then extinguish if no issues are detected. MILs are often combined with a pictorial symbol to help identify the system or component affected.
Insulation Testing	Insulation testing involves the use of a Megohmmeter to conduct a resistance test with a high-voltage power supply provided from a charged capacitor within the unit. You should conduct this test before the system is powered up and measure insulation resistance between the component and vehicle frame or chassis. You must always follow the manufacturer specification for test voltage and results; investigate and resolve any discrepancies before commissioning. In general terms, the higher the resistance result, the greater the insulation.

High-Voltage Component Replacement

Table 4.7 Potential tests to be conducted following the repair or replacement of high-voltage components	
Diagnostic Trouble Codes (DTC)	You should use a scan tool to read and clear diagnostic trouble codes, following system component repair or replacement. You need to conduct a global scan that tests all vehicle systems, as seemingly unrelated issues in any managed system could create or affect the high-voltage operation of an electric vehicle. It is good practice to fully test and scan the vehicle following maintenance or repairs, before and after road testing, to ensure no issues have occurred since driving.
Live Data	Live data, also known as parameter identifiers (PID), is accessed using a diagnostic scan tool that reads serial data from an onboard diagnostic socket or data link connector (DLC). This data will show the information that the electronic control unit (ECU) receives from vehicle components and systems in real time. It is worthwhile remembering that live data is essentially second-hand information as it's the ECU's interpretation of system component operation, which may differ from actual measured values.
Milliohm Testing	A milliohm meter is an electrical diagnostic tool that measures very small amounts of electrical resistance. It is an essential tool for assessing the condition of phase windings within motor generator units or equipotential bonding (EPB). Heat will affect the results of a resistance test using a milliohm meter, so you need to ensure a consistent temperature while conducting a test or use a unit with a temperature compensation capability.

High-Voltage Components

Contactors and System Main Relays (SMR)

Contactors and relays are used throughout the design of electric vehicles. They turn on and off high-voltage circuits within the vehicle systems as a form of heavy-duty electromagnetic switch controlled by the 12-volt auxiliary system. Although the construction of contactors and relays is fundamentally different, the terminology is often freely interchanged when these components are discussed. Sometimes referred to as system main relays (SMR), these switches connect and disconnect the high-voltage battery when the vehicle is turned on and off. Contactors or relays are also used in other high-voltage systems such as vehicle charging and air conditioning.

Figure 4.14 HV Contactors and SMR

High-Voltage Component Replacement

Cable and Wiring, Including Insulated Return

Low-Voltage Cabling

The low-voltage system uses insulated copper wiring to transport electricity around the vehicle to where it is needed. Thin strands of copper are bundled together and coated with a plastic shield to provide insulation and prevent electricity from conducting to any other metal components creating a short circuit.

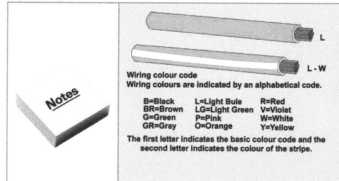

In the low-voltage electrical system, the external plastic coating is usually colour-coded. You can use these colours to help trace cable routing or identify them on a wiring diagram when diagnosing an electrical circuit fault.

This image provides an example, however, always refer to manufacturers naming conventions to avoid misinterpretation.

Low-voltage electrical wires come in different sizes. Copper strands are bundled together, allowing current to still flow even if one or more strands are damaged. Automotive low-voltage wires are usually labelled with the number of strands they contain and the diameter of each strand in millimetres. This indicates the amount of current the wire can carry.

Table 4.8 shows some typical low-voltage wire size designations and potential uses.

Table 4.8 Low-voltage wire sizes and potential uses		
Number of strands / Wire diameter	Continuous current rating	Uses of the wire
9 / 0.30mm	5.75 amps	Side lamps, tail lamps, reversing lamps, horns.
14 / 0.30mm	8.75 amps	Side lamps, tail lamps, reversing lamps, horns, general wiring.
28 / 0.30mm	17.5 amps	Headlamps, fog/driving lamps, windscreen wiper motor.
44 / 0.30mm	27.5 amps	Charging cable (after DC-to-DC converter), auxiliary battery feed.
65 / 0.30mm	35 amps	Charging cable (after DC-to-DC converter).
84 / 0.30mm	42 amps	Charging cable (after DC-to-DC converter).
97 / 0.30mm	50 amps	Heavy duty charging cable (after DC-to-DC converter).
120 / 0.30mm	60 amps	Heavy duty charging cable (after DC-to-DC converter).
80 / 0.40mm	70 amps	Heavy duty charging cable (after DC-to-DC converter).
37 / 0.71mm	105 amps	Emergency starter/Battery cable.
37 / 0.90mm	170 amps	Emergency starter/Battery cable.
61 / 0.90mm	300 amps	Emergency starter/Battery cable.

High-Voltage Component Replacement

The thicker the wire (cross-sectional diameter), the more electricity it can carry and the lower its internal resistance. This means that the component will receive more of the voltage and current. The longer the wire, the higher its resistance. This means that the component will receive less of the voltage and current.

> Doubling the length of the wire doubles its resistance.
> Doubling the diameter halves its resistance.

Terminals, Connectors and Continuity

Vehicle manufacturers design the electrical wiring to create the circuits as insulated sections called wiring looms. These looms can be routed efficiently and hidden from view behind panels, carpets and trims when the vehicle is assembled. The wiring looms are made in sections and joined together by **connectors**. **Terminals** are used to connect the wires to electrical components at the ends of the looms. The circuits must be continuous and unbroken for electricity to operate the components correctly – this is called electrical **continuity**.

Connector - a component that joins two parts of a circuit together.

Terminal - where the circuit ends (terminates).

Continuity - an electrical circuit that conducts electricity and is unbroken (continuous).

High-Voltage Cabling

The high-voltage system in hybrid and electric vehicles requires special designs and construction for the cables and wiring, considering the following factors:

> Large power transmission between the high-voltage battery, inverter, and motor generator.
> Shielding against the electromagnetic interference caused by the power electronics.
> Large differences in temperature due to ambient conditions and electrical loading.
> Safety, armouring, and colour coding of the external insulation.
> Flexibility due to tight installation space.

Figure 4.15 High-Voltage Cables

High-Voltage Component Replacement

The cabling for the power transmission between the battery, inverter, and motor generator unit has a large diameter, which can handle the amount of current required by the system. It can be single or multi-cored and made from copper or aluminium. The design of the core conductor affects the cable's flexibility for use in tight installation spaces.

The cabling reduces the **electromagnetic interference** by using a **shielding** made from **braided** metal strands that surround the conductor, between insulation layers. This shielding also provides a secondary function as **equipotential bonding (EPB)**, which seeks to balance the voltage between different components should a breach of insulation occur. This is a safety system designed to reduce the possibility of electric shock.

The materials used in the high-voltage wiring can operate in temperature ranges between -40° C and 200° C, which account for weather and heat created by high current draw during electrical operation.

The external insulation of the high-voltage wiring is often reinforced to prevent accidental cutting and is coloured bright orange as a warning that it contains high-voltage electricity. The terminals of high-voltage cabling are often of a different size to low-voltage systems to avoid accidental connection to incorrect circuits which may cause damage or injury.

Remember that voltage is dangerous.
It only takes a very small amount of electric current to cause electrocution and death once the voltage touch threshold has been passed. Regardless of the size of the wire or electrical connector, if it is colour-coded orange, it has the potential to carry high-voltage and therefore cause injury or death. Always wear high-voltage personal protective equipment (PPE) when working on or around any wires or connectors coloured bright orange.

Insulated Return

The earth return system in a vehicle poses certain safety issues. In a high-voltage system, the vehicle chassis or frame could become live, causing electric shock and even death, if it was used as part of the electrical circuit. Therefore, the high-voltage systems of electric drive vehicles have an insulated earth return system. The high-voltage wiring is fully insulated from the vehicle chassis and body, and one main cable returns to the negative side of the battery.

Electromagnetic interference - also known as radio-frequency interference (RFI), is a disturbance that affects an electrical circuit due to electromagnetic induction.

Shielding - the practice of reducing or redirecting the electromagnetic field (EMF) in a space using barriers made of conductive or magnetic materials.

Braided - a type of electrical conductor made from multiple strands of thin wire woven together.

Equipotential bonding (EPB) - an electrical safety measure that involves connecting various exposed conductive parts to maintain them at substantially the same voltage potential.

High-Voltage Component Replacement

Motor Generators

Motor generators are essential for the propulsion and power system of hybrid and electric vehicles. They use magnetism to either provide motion or electric current. A motor turns mechanically when it receives electric current. A generator creates electric current when it turns mechanically. Hybrid and electric vehicles use different types of motor generators.

Direct Current Motors

A simple direct current (DC) motor can be made by passing an electric current through a coiled wire that is wound around a central shaft called the armature. This creates an electromagnet.

> The electric current produces an invisible magnetic field, which is repelled or attracted by the permanent magnets surrounding it. This makes the **armature** turn.
> The armature would normally stop once it aligns with the magnetic field.
> To keep it rotating, the **polarity** of the electricity passing through the electromagnet on the armature must be changed. A component called a **commutator** does this.
> Two spring-loaded electrical contacts called **brushes** are mounted on the end of the armature. They maintain an electrical connection with the commutator as the shaft rotates.
> As a new section of the commutator aligns with the brushes due to the revolving of the armature, polarity is swapped, and the motor continues to rotate.
> The motor stops when the electric current is switched off.

The motor produces more power when the magnetic forces inside it are stronger. The magnets inside the motor casing that surround the armature can create stronger magnetic fields if they have wire coils wrapped around them and an electric current is circulated. These external magnet wires are called **field coils**.

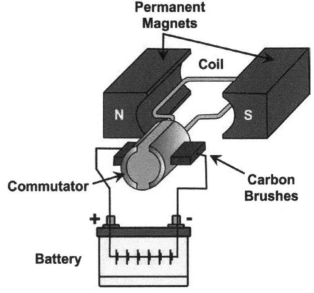

Figure 4.16 Direct Current Motors

Armature - the rotating shaft of an electric motor.

Polarity - the positive and negative terminals of an electric circuit or the north and south poles of a magnet.

Commutator - a segmented electrical contact on the end of a motor armature, designed to switch electrical polarity as the motor rotates.

Brushes - spring-loaded electrical contacts that transfer current to the commutator.

Field coil - copper wiring around magnets that can increase the magnetic field when electrified.

High-Voltage Component Replacement

Direct Current Brushless Motors

A direct current brushless motor avoids some of the problems encountered in a standard DC motor, where electric current has to be supplied to a rotating armature. Three sets of **field effect** electromagnets are arranged in a **stator** and a permanent magnet rotor is mounted within their magnetic field. A control unit delivers a pulsed DC input through three feed circuits that switch from positive to negative between a pair of **electromagnets** opposite each other. The resulting magnetic field attracts the rotor and aligns it within the energised field. If the next pair of electromagnets is then energised in the direction of rotation, the rotor moves to align within their field. If the control unit repeats this process to each of the electromagnetic pairs in turn, rotation is created. The control of the DC supply to each of the three electromagnetic pairs comes from a **pulse width modulated** signal and the **frequency** determines the rotational speed.

Field effect - a phenomenon in which the electrical conductivity of a material is changed by applying an external electric field.

Stator - the stationary component of a motor or generator.

Electromagnet - a device that produces a magnetic field by passing an electric current through a coil of wire wrapped around a core of magnetic material.

Pulse width modulation (PWM) - a technique used to control the average power or amplitude of an electrical signal by changing its pulse width or duration.

Frequency - how often something occurs.

Alternating Current Motors (AC)

The alternating current (AC) motor is another type of electric motor that uses magnetic fields. Unlike a direct current (DC) motor, where polarity has to be continuously changed by a commutator, alternating current changes direction naturally as it operates, and this can swap polarity and keep the motor rotating.

AC motors can be divided into different designs, including:
- Brush type
- Synchronous type
- Induction type
- Three phase type
- Axial Flux

Brush Type AC Motors

A simple brush type AC motor design uses two permanent magnets placed on either side of a rotating wire coil, similar to the DC motor. The magnets are aligned so that the coil faces the north pole of one magnet and the south pole of the other. Conducting brushes touch two slip ring connectors, which feed the inner coil of the armature with electric current. When the current flows, magnetic forces make the armature coil rotate so that the south pole turns to the north pole of the permanent magnet, and vice versa. When the supply current alternates, the magnetic poles of the coil swap places, and the motor keeps turning. The frequency of the alternating current affects the motor speed.

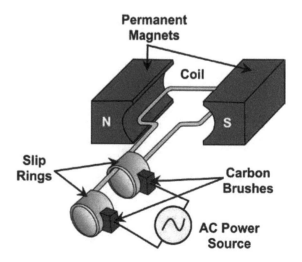

Figure 4.17 Brush Type AC Motors

High-Voltage Component Replacement

Synchronous Type AC Motors

A synchronous type AC motor produces a very precise motor speed. This design has a set of coils surrounding a rotor, but the rotor consists of permanent magnets instead of a wire coil. The electric coils are arranged as opposing pairs in a stationary housing around the edge of the motor, known as the stator. The north-south pairs attract the north-south poles of the permanent magnets in the rotor, making it turn. As the alternating current cycles back and forth, the motor rotates with a very accurate speed that depends on the current frequency and the number of coils used. This type of system creates a lot of heat when used in hybrid and electric vehicles, which often require a dedicated cooling system.

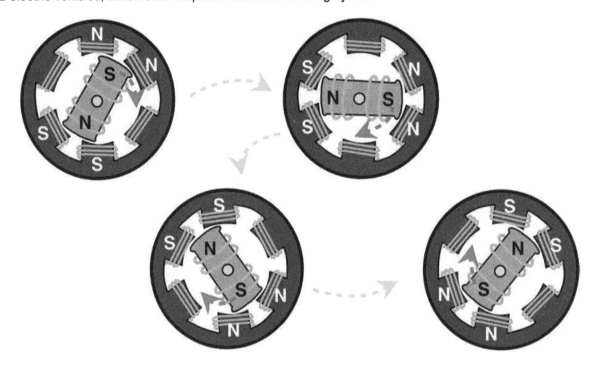

Figure 4.18 Synchronous Type AC Motors

The permanent magnets used in electric vehicle motors are very strong "rare earth" magnets. Despite their name, rare earth magnets are made from fairly common compounds, such as neodymium iron boron (NIB) or samarium cobalt. These rare earth magnets are about 10 to 15 times stronger than a standard iron magnet, making them ideal for powerful electric motors.

Rare earth magnets that are used in vehicle electric motors have an extremely strong magnetic attraction to each other or metal surfaces.

- They can cause injuries to body parts pinched between two magnets, or a magnet and a metal surface, such as broken bones or severed fingers.
- NIB magnets are very brittle and can chip and shatter if they get too close to each other. The flying chips can cause injuries.
- The magnetic fields of rare earth magnets can disrupt the operation of electronic life sustaining equipment, such as heart pacemakers.

You must always be careful when working on or around these magnets, and follow the manufacturer's instructions, including the use of any specialist tools.

High-Voltage Component Replacement

Induction Type AC Motors

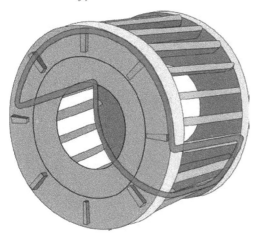

An induction type AC motor does not use brushes or conducting slip rings. Instead, it uses an effect called induction, where a changing magnetic field induces an electric current in a similar way to a transformer coil. A series of conductor coils around the edge of the motor produce an invisible magnetic field when an alternating current passes through them. This magnetic field induces a current in a winding attached to the armature of the motor, creating its own magnetic field, making it turn. This type of motor avoids the wear or sparking caused by brush type motors.

Figure 4.19 Induction Type AC Motors

 Because the inner conductor coils on the armature of an induction type motor have a round, cage-like shape, engineers often call this design a 'squirrel-cage' motor.

Three Phase Type AC Motors

Many hybrid and electric vehicles use an **inverter**, which converts the high-voltage DC stored in the batteries into an AC current in three separate **phases**. Each phase is arranged 120 degrees from the next, forming a complete 360-degree cycle. Powerful electric traction motors are often wired for this three-phase electricity. These motors have coils spaced 120 degrees apart, each driven by one phase of the electricity. This arrangement creates a rotating magnetic field simply and efficiently, which turns a permanent magnet rotor. An advantage of this design is that it can often be air cooled, reducing weight and cost.

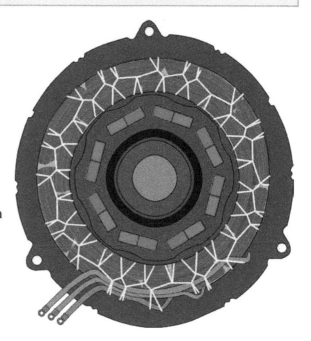

Figure 4.20 Three Phase Type AC Motors

 **Inverter** - an electronic component that converts direct current (DC) into alternating current (AC).

Phase - the difference in timing between each cycle of different alternating currents.

High-Voltage Component Replacement

Axial Flux Motors

An axial flux motor, sometimes called a pancake motor, is one in which two magnetic rotors are placed on either side of the stator. This is different from radial flux motors, where the rotor is placed inside the stator.

The rotors, which are the part that spins, have magnets arranged in a circular pattern along their axis (on its side). The stator, which is the stationary part, has coils wound around iron cores, which create a magnetic flux that surrounds the rotors. When an alternating current (AC) is applied to the coils, the magnetic field interacts with the magnets on the rotor, causing it to rotate.

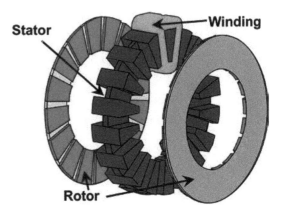

Figure 4.21 Axial Flux Motors

Some of the advantages of axial flux motors used in the design of electric vehicles include:
- High power density: The axial flux design allows for a smaller and lighter motor, which saves space and weight in the vehicle.
- High efficiency: The axial flux design reduces the air gap between the rotor and the stator, which minimises the magnetic losses and increases the power output.
- Low noise: The axial flux design reduces the mechanical friction and vibration of the motor, which lowers the noise level and improves the comfort of the vehicle.

Some of the disadvantages of axial flux motors used in the design of electric vehicles include:
- High cost: The axial flux design requires more magnets and coils, which increases the material and manufacturing costs of the motor.
- Thermal management: The axial flux design generates more heat in the motor than equivalent radial motors, which requires a cooling system to prevent overheating and damage.
- Control complexity: The axial flux design requires a sophisticated controller to regulate the current and voltage of the motor, which increases the software and hardware complexity of the vehicle.

 Alternating current (AC) motors provide more flexibility than direct current (DC) motors in the propulsion design of electric vehicles. With a DC motor, if voltage is increased, so is its speed. However, with an AC motor, frequency increases speed, and voltage increases power.

Inverter Converters

Most electric drive motors used by hybrid or electric vehicles operate using high-voltage, three phase alternating current (AC). This means that a device is needed to convert the high-voltage DC stored in the battery unit into high-voltage AC. This is the job of the inverter.

DC battery voltage is first fed to a set of **capacitors** inside the inverter. The capacitors act as a temporary storage device and buffer between the high-voltage vehicle battery and the electric drive motors.

The inverter then converts the direct current (DC) into alternating current (AC) using a series of electronic switches called insulated gate bipolar **transistors** (IGBTs). The IGBTs can handle large amounts of power with very fast switching rates.

The transistors are arranged in pairs, which direct electric current in either one direction or another when activated. By switching alternating pairs, the current changes direction, turning DC into AC [see **Figures 4.22** and **4.23**]. If three sets of transistors are used, the system can supply three phase electric current with varying frequency and voltage. The IGBTs have such a fast-switching rate that they produce a very smooth AC wave, which the AC vehicle drive motors can use effectively.

High-Voltage Component Replacement

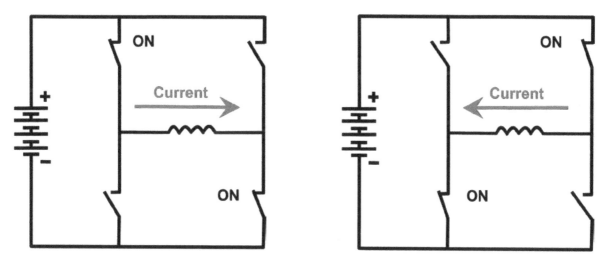

Figure 4.22 Current Direction Swapped Using Switches

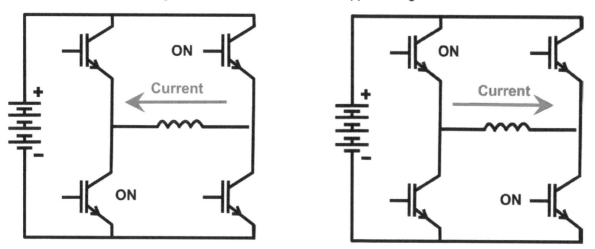

Figure 4.23 Current Direction Swapped Using Transistors

When motors are mechanically driven, they become electric generators, and the energy can be used to recharge the high-voltage battery. The energy created in the generators is an alternating current (AC). This means that the electricity must be converted to direct current (DC) before it can be used in the battery. The process of converting AC to DC is known as **rectification**. Rectification in the inverter converter is achieved using **diodes** to redirect alternating current so that it all leaves on the same route as direct current, regardless of which direction it arrives at the converter. Before returning to the high-voltage battery, the rectified electricity is first directed through the capacitors, where any minor AC ripples are smoothed out. This ensures that an effective DC supply is provided to the high-voltage system.

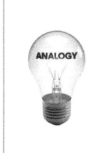

ANALOGY

The process of rectification can be likened to traffic control at a junction. If correctly arranged, lane control can be used so that, no matter which direction vehicles arrive at the junction, they are all required to leave via one dedicated road. In this analogy, diodes, which can be considered one-way valves for electricity, can be used as the method of traffic control to redirect the arriving AC electricity into DC.

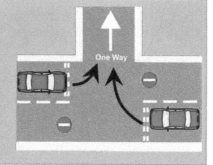

High-Voltage Component Replacement

The capacitors used inside the inverter converter are able to store electrical energy for a considerable amount of time. This electrical energy will be higher than the voltage touch threshold for dry human skin and therefore could cause electrocution, injury and death if not correctly discharged. The inverter converter will have methods for both actively and passively discharging the capacitors; however, this may take several minutes, and absence of voltage must be correctly assessed wearing high-voltage personal protective equipment (PPE) and using correctly rated and calibrated test equipment before any work is started.

Capacitor - a device that temporarily stores electrical energy in an electric field.

Rectification - the process of converting alternating current (AC) to direct current (DC).

Transistor - an electronic semiconductor component that acts as an electronic switch with no moving parts.

Diode - an electronic semiconductor component that acts as a one-way valve for electricity.

DC-to-DC Converters

The low-voltage auxiliary system of an electric vehicle is supplied with energy from a step-down transformer inside a component called a DC-to-DC converter. The converter uses direct current (DC) electricity from the high-voltage battery, which it first inverts to high-voltage alternating current (AC) inside its transformer coils. The transformer then uses the oscillating magnetic field created by the high-voltage alternating current to induce a low-voltage alternating current in a step-down coil. The low-voltage alternating current is then rectified into low-voltage direct current, which can be used to charge the auxiliary battery and supply power to all low-voltage auxiliary systems.

Figure 4.24 A DC-to-DC Converter

The principal job of the low-voltage auxiliary battery on an electric vehicle is to maintain critical systems such as computer memory, central locking, and security while the vehicle is switched off, and supply the required energy needed to switch on the high-voltage system via contactors and system main relays. Milliseconds after the high-voltage system has been energised, the low-voltage auxiliary load is transferred to the supply from the DC-to-DC converter. This means that the auxiliary battery will often be a small low-capacity unit, as it doesn't need to supply all of the auxiliary electrics.

Although titled a DC-to-DC converter, as the component uses a transformer, AC is required as part of the step-down process. The terminology DC-to-DC converter is intended as a simplified summary description of the overall process.
The DC-to-DC converter may be combined inside the inverter converter casing, but more often will be a stand-alone component with its own individual unit, mounted as part of the auxiliary system.

High-Voltage Component Replacement

High-Voltage Air Conditioning Compressors

Hybrid and electric vehicles use a refrigerant compressor in their air conditioning and climate control systems driven by an alternating current (AC) electric motor from the high-voltage system. The compressor can be direct current (DC) fed with an onboard inverter, or AC fed from the vehicle's main inverter unit. These compressors require a special Polyalkylene Glycol (PAG) or Polyolester Oil (POE) oil in order to function safely. The oil should not readily conduct electricity, as this may cause a short circuit in the high-voltage system, making metal components become 'live'. Once opened, the hygroscopic nature of PAG and POE oils means that any moisture absorbed will increase its electrical conductivity. This means they don't have a storage 'shelf-life'; any oil left over following replenishment of the recovery machine should be disposed of following safety and environmental requirements. Always follow manufacturers' instructions. For more detailed information on the operation and maintenance of air conditioning, [see Chapter 3].

Figure 4.25 An HV AC Compressor

The number of high-voltage wires feeding the air conditioning compressor may give an indication of whether it is direct current (DC) or alternating current (AC) powered. Two wires often indicate DC, and three wires indicate AC.

Many compressor manufacturers use a form of PAG oil as the recommended original equipment manufacturer (OEM) lubricant.
Using a replacement oil other than that recommended by the manufacturer may invalidate any potential warranty claim due to the failure of a compressor. Always use the manufacturer-specific recommended oils for maintenance and repair when a vehicle compressor is within its warranty period.

Battery Management Systems (BMS)

The battery management system (BMS) is an electronic control unit that has the main purpose of maintaining the high-voltage battery. Without some form of management, the degradation of the battery would be such that its state of health (SoH) would decline in a relatively short period of time. Its main functions are to monitor the state of charge (SoC) and the temperature and to provide output control over thermal management, charge/discharge cycles in conjunction with the electric vehicle control unit and the cell balance system.
Thermal management normally involves cooling the high-voltage battery, as a byproduct of the charge/discharge process is heat. However, it can also preheat batteries in colder climates. Thermal management will often make use of airflow, liquid coolant, or air conditioning to maintain heat within a range specified by the manufacturer, or electric heating elements for pre-conditioning of the modules.
The state of charge is controlled by balancing. Depending on the manufacturer's preference, balancing may be active or passive. For more information on the process of battery balancing, [see Chapter 5].

Figure 4.26 A BMS Unit

High-Voltage Component Replacement

PTC Heaters

Electric vehicles and some hybrids require a method of heating the passenger compartment as they may not have the provision for using engine-heated coolant and heater matrix to supply this function. Instead, they are able to use a positive temperature coefficient (PTC) thermal resistor as an electric heating element to warm air or liquid, which can then be used to heat the cabin. Heating of the cabin can sometimes be used to indirectly warm and condition the high-voltage drive batteries.

Figure 4.27 A PTC Heater

As many heating elements are powered from the high-voltage drive battery and this can be a drain that may affect the available range, some manufacturers provide alternative methods to keep occupants warm. Powered from the low-voltage auxiliary system, electrically heated seats and steering wheel can reduce the load on the high-voltage drive battery and provide passenger comfort.

A positive temperature coefficient PTC heater is a form of thermal resistor which converts electrical energy into heat. A good analogy is that of an electric kettle. When filled with cold water, the electrical element has a relatively low resistance. When switched on electricity flowing in the element is converted into heat by the resistance, warming the water. Heat and electrical resistance are very closely linked. As electric current flows in the kettle's element, and is converted into heat, the heat in turn creates more electrical resistance and the process grows exponentially.

Hydrogen

Hydrogen is the most abundant element in the known universe, and it makes an extremely good energy carrier for operating electric vehicles. Hydrogen is extremely flammable but produces very little in the way of harmful emissions (the main by-product being water). Unfortunately, hydrogen does not occur naturally on Earth, so it must be manufactured.

The process of making hydrogen is fairly straightforward. It normally involves separating the hydrogen from other elements in compounds by a process of **electrolysis**. For example, separating the hydrogen from oxygen in water by passing an electric current through it.

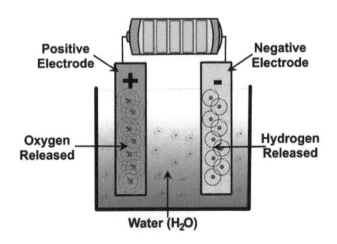

Figure 4.28 The Electrolysis of Water

High-Voltage Component Replacement

In general terms, it takes around three times as much energy to separate the hydrogen as can be obtained from the hydrogen itself. This makes hydrogen an inefficient fuel source in many ways.

Also, hydrogen **molecules** are so small that they will naturally leak through almost any container. This means that storage can be a problem. For example, if you filled up a standard metal or plastic fuel tank with hydrogen, even if you didn't use the vehicle, the tank would be empty in a few days.

Electrolysis - a chemical decomposition produced by passing an electric current through a liquid or a solution containing ions.

Molecules - the smallest component of a chemical compound that can take part in a chemical reaction.

Hydrogen Fuel Cell

A method that can be used to power vehicles using hydrogen, other than burning it inside an engine, is a fuel cell. Some manufacturers produce vehicles with a hydrogen fuel cell, but because of their complexity they are often more expensive than an equivalent battery electric vehicle.

The fuel cell uses hydrogen to create electricity that can then be used to power electric motors.
A standard vehicle battery, including those found in hybrid and electric vehicles, stores all of its chemical energy inside a casing which it uses to create electricity through a chemical reaction. Once the battery has used up all of its chemical energy, the battery is considered flat. Vehicle batteries can reverse the discharge process by supplying an electric current with a voltage potential higher than that coming out of the battery. This is normally achieved using a generator or by plugging into an electricity outlet.

A fuel cell is similar to a battery, but it doesn't store its own internal electricity in the form of a charge. With a fuel cell, as long as the cell is kept supplied with chemical elements, in this case hydrogen and oxygen, then it works like a battery that never goes flat. Hydrogen is stored in a separate container/fuel tank, and then mixed with the oxygen inside the fuel cell to create electricity.

The hydrogen fuel cell is not a new idea. Sir William Grove invented the first fuel cell in 1839. He knew that water could be separated into hydrogen and oxygen if an electric current was passed through it by a process known as electrolysis.
Grove found that if the process was reversed, he could create electricity by recombining the hydrogen and oxygen and producing water. He then went on to create a very basic type of fuel cell, which he called a gas voltaic battery.

Fuel Cell Construction

The most common type of hydrogen fuel cell is made using a component called a proton exchange membrane (PEM) which acts as a **catalyst**. This is a material that separates the two sides of the fuel cell.
- One side of the fuel cell is fed with oxygen from the surrounding air.
- The other side is fed with hydrogen from a pressurised fuel tank.

Catalyst - A component that starts and maintains a chemical reaction but is unaffected by the process.

High-Voltage Component Replacement

Fuel Cell Operation

As hydrogen enters the fuel cell, a catalytic reaction takes place when it contacts the PEM. This reaction strips the **protons** from the hydrogen atoms and moves them through the **membrane** towards the oxygen on the other side. This leaves the **electrons** from the hydrogen atoms, which travel through a different circuit and create electric current. After the energy from the hydrogen has been converted into current and powered the vehicle's electric circuit, the electrons reattach themselves to the protons of the hydrogen atoms and combine with the oxygen to form water (H_2O). This means that the only emissions from the fuel cell are water and heat, making it clean and non-polluting at point of use.

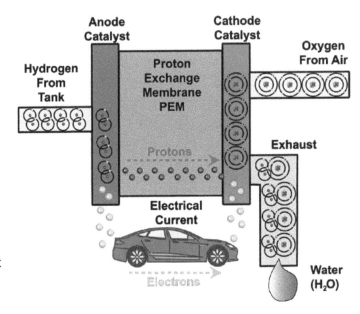

Figure 4.29 Fuel Cell Operation

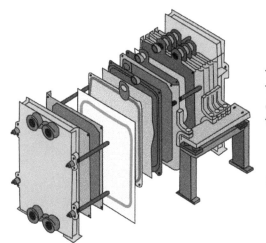

The typical output from a single fuel cell is approximately 0.8V. This means that a number of fuel cells have to be combined (known as a fuel cell stack) to create a usable amount of voltage to drive electric motors.

Figure 4.30 A Fuel Cell Stack

Hydrogen Storage

A hydrogen storage tank is often made from carbon fibre and pressurised to around 700 bar in order to carry sufficient quantities needed to power an electric vehicle for an acceptable mileage range. These tanks can be filled at a fuel station in a similar manner to that used to fill a vehicle with petrol or diesel. However, a pressurised nozzle is sealed and locked into the tank during the filling operation. In order to supply enough pressure to fill the hydrogen tank, the fuelling rig will often operate at a pressure of around 900 bar.

This means that drivers and operators can refuel in a similar manner to that used with traditional petrol and diesel and then drive away on electricity.

Protons - the positively charged particles of an atom.

Membrane - a thin layer of material that is used to separate two connected areas.

Electrons - the negatively charged particles of an atom.

High-Voltage Component Replacement

Electric Vehicle Charging Systems

Unless the electric vehicle is some type of hybrid or is able to carry a supply of energy in the form of hydrogen using a fuel cell stack, the high-voltage drive batteries will require charging, which will normally come from the mains electricity grid.

Generated mains electricity is supplied in the form of alternating current (AC), and chemical storage in a battery requires direct current (DC). As a result, if an AC supply is used, it will require conversion via a rectifier before it can charge the drive battery. This is achieved by the use of an onboard charger (OBC). The purpose of the onboard charger is to work in conjunction with the battery management unit and control any required charging from the mains electricity grid.

Two key terms are used to explain external charging: Mode and Type.

Put simply:
- Mode describes the method of charging, for example, slow, fast, rapid, AC, or DC.
- Type refers to the shape of the plug and socket used for connection from the charging grid to the vehicle.

The owner of a plug-in or all electric vehicle may need to meet certain criteria in order for it to be a viable option. Do they have access to off-road parking, such as a driveway or garage, where the vehicle can be re-charged overnight, for example?

Unless official permission is granted by the local authority, it is not advisable to trail an electric cable across pavements or other public areas to connect a plug-in electric car parked on-street with a private/household electricity supply.

Modes of charging are normally categorised into four styles.

There will be subdivisions in the following descriptions, however, voltages, amperages and charge times will not be stated, because as the charging technology develops, any figures quoted will be rapidly outdated.

Mode 1 - Single or three-phase AC charging
Mode 1 is a charging method mostly used with early electric vehicles. It is a simple extension lead that could be plugged into a standard single phase domestic electricity outlet or three phase commercial supply. There is no intelligent communication with the vehicle, and the only protection is the fuse in the mains plug itself or a circuit breaker. As a result, this mode of charging is mostly obsolete.

Mode 2 - Single or three-phase AC charging
Mode 2 is an upgraded version of mode 1, which can be used with a single-phase domestic electricity outlet or specialised three-phase connector. Unlike mode 1, however, it is supplied with an intelligent in-cable charging box, known as an ICCB. The ICCB allows the vehicle to communicate with the charging lead and regulate its operation as required. Due to the very slow charging style, mode 2 is sometimes nicknamed 'granny charging'.

Mode 3 - Single or three-phase AC charging
Mode 3 refers to any purpose-built/dedicated AC charging unit. Depending on its location and connection to the power grid, the capability for single or three-phase can be utilised. This means that dedicated equipment fitted to domestic residences will normally only have the capability to charge using the available single phase, however, if the equipment is fitted in a commercial environment, it may be able to deliver a three-phase charge. This gives a range of charging speeds from slow to fast. Mode 3 equipment may be tethered or untethered, referring to whether a charging lead is permanently fixed to the charging equipment. The advantage of using untethered equipment is that the vehicle owner may use an adapter to change the plug type (style), increasing the range of vehicles that can be charged from a single unit.

High-Voltage Component Replacement

Mode 4 - DC rapid charging
Mode 4 is a dedicated DC charging station, providing a rapid charge that can bypass most of the vehicle's onboard charging unit, supplying electrical energy directly to the high-voltage drive battery.
It achieves this by the charging station being connected to a commercial AC supply and performing the rectification process inside the unit (off-board) and then connecting to the vehicle via a specialist DC cable and socket.

Charge speeds are often described by their power delivery, for example:
- Slow - 2.4 to 6 kW
- Fast - 7 to 22 kW
- Rapid - 25 to 100 kW
- Ultra rapid - 100 to 350 kW

Regardless of the method or mode of charging chosen, the onboard charger fitted to the vehicle will set the capable limits that the particular high-voltage drive battery can charge at. This may be dictated by cost, limitations of the vehicle, or a design feature to protect the health and longevity of the battery.
This means that although a vehicle may be plugged into a rapid high energy capable charger, it will only charge at the maximum rate and speed set by the vehicle manufacturer.

Tesla Supercharger and Destination Charger

Outside of the general description of Mode 1 to 4 charging, Tesla has a network of rapid DC charging stations, known as Superchargers. Designed primarily for use with Tesla electric cars, it is a 480 volt system with its own style of connector, which is able to deliver a range of electrical power supplies. The company also has slower charging stations, known as Destination chargers, at locations which are often used for purposes other than specifically charging, such as restaurants or hotels.

When a vehicle is plugged into a charger, although many safety systems are employed to reduce the possibility of electrocution to the operator, by its very nature, the high-voltage system cannot be shut down and is therefore live. Furthermore, the electrical grounding of the charging station, although designed for electrical safety, provides an alternative pathway for electrical current, meaning that if someone touches the vehicle or charger that has a breach of insulation, the risk of electrocution is increased. Because of this, if a vehicle is placed on charge in a workshop environment, it should be in a cordoned off area with signs clearly stating that the vehicle is on charge and should not be touched. No physical, electrical, mechanical work or maintenance should be conducted while the vehicle is plugged into a charger. Always disconnect the charger before commencing any work.

Types of Charging Socket

During the production of early electric and plug-in hybrid vehicles, manufacturers had no standard to work with when it came to the design and shape of charging sockets used. As a result, several types of sockets were produced and used across a mixture and range of vehicles. This led to the design and style of charging sockets to become known as 'Types'.

As with most products supplied, if a choice is offered, over time some designs will become more popular than others, leading to a dominant style which is adopted by the majority of manufacturers.

High-Voltage Component Replacement

All designs will consist of connection pins and sockets to allow charging to take place and communication between the electric vehicle supply equipment (EVSE) charging station and the vehicle. The pin and socket configurations normally conform to the following nomenclature:

AC high-voltage charging pins and sockets
- L1 AC phase 1
- L2 AC phase 2
- L3 AC phase 3
- N Neutral
- PE Protective earth

DC high-voltage charging pins and sockets
- \+ Positive DC connection
- \- Negative DC connection

Communication pins and sockets
- PP Proximity Pilot - communicates that connection between the vehicle and charging station has been made, and also supplies information about the cross-sectional area of the charging cable, and therefore its current carrying capabilities.
- CP Control Pilot - connects with the in-vehicle network and allows bi-directional communication between the vehicle and charging station.

Type 1 – Yazaki

Type 1 charging connectors are primarily used by designers and manufacturers in Asia and North America. This style will work with single phase AC electricity only, up to 250 volts. It only has three high-voltage connections, consisting of L1, N, and PE, meaning that regardless of available supply, it can only ever charge with single phase AC. It also contains the communication ports CP (control pilot) and PP (proximity pilot).

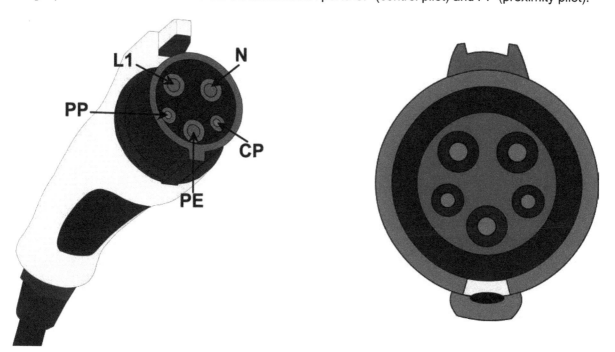

Figure 4.31 A Type 1 Plug and Socket

High-Voltage Component Replacement

Type 2 – Mennekes

Type 2 charging connectors are primarily used by designers and manufacturers in Europe. This design has five high-voltage connections: L1, L2, L3, N, and PE, meaning that it has the flexibility, depending on supply and vehicle capabilities, to charge using either single or three phase electricity. If connected to a single-phase supply, only pins L1, N, and PE are used. However, if three phase is available, it will then connect L2 and L3, increasing the speed of charge available. It also contains the communication ports CP (control pilot) and PP (proximity pilot).

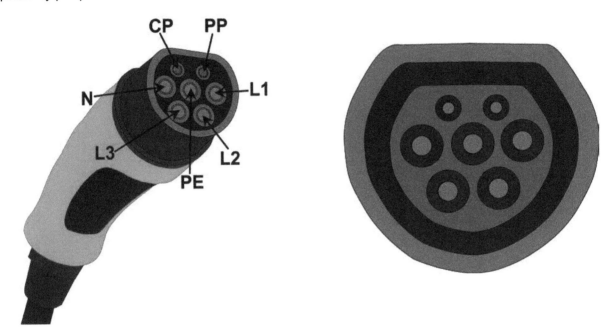

Figure 4.32 A Type 2 Plug and Socket

 Due to the flexibility to make use of either single or three phase electricity supply, Type 2 tends to be the dominant style preferred by manufacturers and vehicle designers.

Type 3 – Scame

Type 3 charging connectors were primarily used by designers and manufacturers in Europe. This design came in two styles: Type 3A for single phase and Type 3C for three phase supplies. Type 3A used pins L1, N, and PE, whereas 3C used L1, L2, L3, N, and PE. Both styles also contained the communication ports CP (control pilot) and PP (proximity pilot).

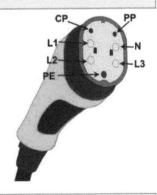

Figure 4.33 A Type 3 Plug

 In January 2013, the Type 2 connector was adopted by the European Commission as the official charging connector, meaning that the Type 3 socket is mostly obsolete.

Charging socket standards can be found in the appendix.

High-Voltage Component Replacement

Type 4 – CHAdeMO

Type 4 charging connectors are primarily used by designers and manufacturers in Asia and North America for dedicated DC rapid charging. It has two DC charging pins: + (positive) and - (negative), and communication pins in a different format than the CP and PP found on AC charging connectors. CHAdeMO is normally associated with vehicles that are supplied with a Type 1 charging socket for AC connection, but as the plug design is fundamentally different from other styles, it requires vehicles using this style to have two charging sockets, often mounted side-by-side.

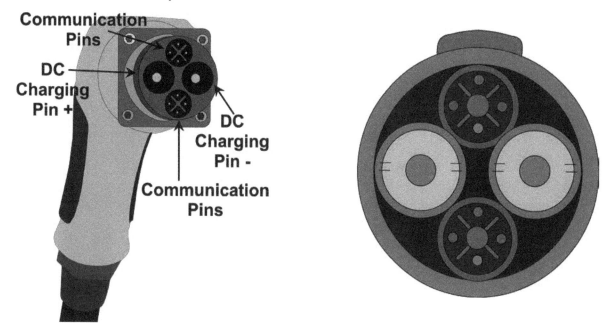

Figure 4.34 A Type 4 Plug and Socket

CCS Combined Charging Systems

To provide manufacturers with the option of using both AC and DC rapid charging with a simple socket design, CCS or Combined Charging Systems are available, where a two pin DC socket is placed below a Type 1 or Type 2 AC socket. This allows the vehicle to be plugged into a Type 1 or 2 when only AC supply is available or use a specialised plug with a combination of a Type 1 or 2 mounting and DC rapid charging pins when DC is available. When used with a DC charging source, the only pins required in the AC charging socket are the protective earth, proximity pilot (PP), and control pilot (CP).

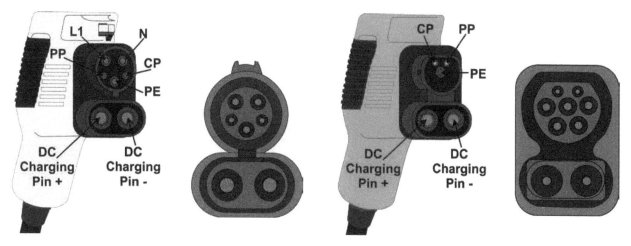

Figure 4.45 CCS Type 1 and 2 Plugs

High-Voltage Component Replacement

NAS North American Standard

NAS charging plugs are a type of charging connector system developed by Tesla. They have been used on all North American market Tesla vehicles since 2012 and were opened for use to other manufacturers in November 2022. Several other vehicle manufacturers have announced that they will adopt the NAS standard for their electric vehicles in North America.

NAS charging plugs have five pins: DC+/L1, DC-/L2, G, CP, and PP. The DC+/L1 and DC-/L2 pins carry the direct current (DC) for fast charging, as well as the alternating current (AC) for normal charging. The G pin is the ground pin for safety. It also contains the communication ports CP (control pilot) and PP (proximity pilot).

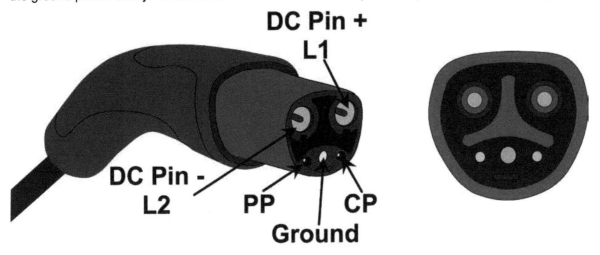

Figure 4.46 A NAS Plug

Charging Communication

When a vehicle is connected to the charging system, bi-directional communication is initiated between the vehicle and electric vehicle service equipment (EVSE) to agree on the charging process. This is achieved using the control pilot (CP) and the proximity pilot (PP).

The pin of the proximity pilot lead has a resistor connected in series, and then the circuit is completed by joining it to the protective earth.

The pin of the proximity pilot, inside the charging socket is sometimes longer than the others, making it the first to connect and the last to disconnect when plugged in. This signals to the charging unit that a connection has been made, and the size of the resistor indicates the cross-sectional area of the charging cable and therefore how much current it is capable of carrying.

Once fully connected, the control pilot will communicate between the vehicle network and the charging station, normally via a modulated pulse signal.

The amplitude of the signal voltage (height) is used to indicate status:
- Vehicle is connected ready to charge 9v
- Vehicle is being charged 6v
- Vehicle is charging with thermal management 3v

The duty cycle percentage of the pulse communication indicates the amount of current used for charging.

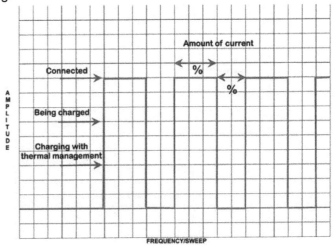

Figure 4.47 Examples of Charging Communication Waveforms

High-Voltage Component Replacement

With DC rapid charging, the communication is normally direct between the EVSE and the vehicle CAN Bus system, as the onboard charger (OBC) is not required for AC to DC current rectification.
Type 4 CHAdeMO has dedicated pins in the charging socket to allow it to communicate with the vehicle CAN Bus.
CCS often uses the control pilot (CP) and protected earth (PE) pins for CAN Bus communication through a process known as powerline communication (PLC).

Wireless Charging

Wireless charging of electric vehicles is achieved through **mutual induction** via a pair of coils, in a similar process to that of a transformer. An **oscillating** electromagnetic field is created in the charging coil, which induces a voltage in the coil of the vehicle's charging circuit. The two coils usually need a fairly precise alignment and proximity, but charging can be obtained with no physical connection between the two. Some of the advantages of wireless charging include:

- No charging cables are required, reducing the hazards created by a trailing wire.
- Easy connection; the driver simply has to position the vehicle in the proximity of the charger. This could be as straightforward as simply parking in a bay and might not require the driver to exit the vehicle.
- As technology improves, it offers the opportunity to develop **infrastructure** to increase the availability of the charging network.

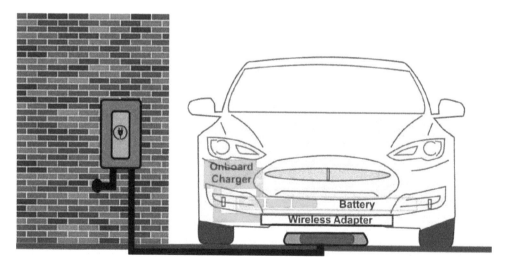

Figure 4.48 Wireless Charging

Mutual induction - a principle in electromagnetism where the change in current in one coil induces a voltage in a neighbouring coil.

Oscillating - the motion of moving back and forth in a regular rhythm.

Infrastructure - the fundamental facilities and systems serving a country, city, or area.

High-Voltage Component Replacement

Vehicle to Grid (V2G)

One of the biggest problems facing electric vehicles is the charging network and infrastructure. There is a well-established worldwide supply chain for fossil fuels used in internal combustion engines, and even in remote locations, the portability of petrol and diesel means that access is often available regardless of local infrastructure. The transition to electric vehicles presents many issues relating to the availability of charge points, as well as the amount of electricity required to service the demands of the owners and drivers of electric vehicles.

Mechanically generated electricity is produced in the form of alternating current (AC), which cannot be stored in that format for use at later times, and therefore it is often created on demand. Chemical storage of energy in batteries requires that the AC is inverted to direct current (DC), and this will produce some energy loss during the process, making it inefficient at best.

The mechanically generated electricity grid needs to be monitored in real time, and during periods of high demand, this often puts a strain on supply. Due to the lack of general electricity storage capacity, not all available sources are used, especially during times of low demand, and many generation opportunities, especially from green low carbon supply, are missed.

Vehicle to grid, also known as V2G, is a bi-directional system that allows electric vehicles to not only charge their batteries from the mains supply network, but also to serve as a supplemental electricity supply source during periods of high demand. With only a small number of vehicles using V2G, this means that batteries may see a significant draw and reduce overall usable capacity and therefore driving range. However, if a large number of V2G vehicles are connected, this becomes a significant electrical storage facility. This would provide the ability to generate excess electricity during periods of low demand, especially from green sources, and divide it up between the V2G network. Then, during periods of high demand, this could be returned to the grid with little impact overall on individual vehicles.

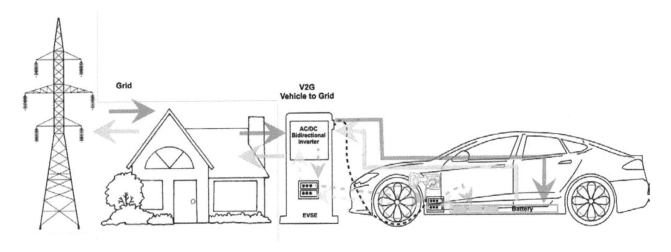

Figure 4.49 Vehicle to Grid V2G

Summary

This chapter has described:

- Safety precautions and procedures when working on EV/Hybrid vehicles.
- Battery technology.
- Removing/refitting and testing high-voltage components.
- An overview of high-voltage components.
- Electric vehicle charging systems.

Chapter 5 High-Voltage Diagnosis and Testing

In this chapter, you will learn about the diagnosis and testing of high-voltage electric vehicle components and systems. Although it is not possible to describe all systems and components, you will be provided with an overview of several of the most common diagnostic test routines. To understand many of the tests, it is expected that you have sufficient prior knowledge of automotive electrics and diagnostic procedures. Some of the test techniques will involve transferable skills which could be utilised when conducting diagnosis on other components and systems and may be suitable for both high and low voltage applications.

In order to effectively test and diagnose electrical faults, it is often necessary to have the circuit powered up and 'live'. This presents additional risks and hazards for the vehicle, but also increases the dangers for those conducting work. Any diagnosis and testing must always be conducted in conjunction with the manufacturer's specific procedures and data, following all high-voltage safety precautions including personal protective equipment (PPE). No 'live' testing or diagnosis should be conducted unless you have received sufficient training and have appropriate experience.

Contents

- Information sources **Page 156**
- Safety precautions and procedures when working on EV/Hybrid vehicles **Page 156**
- High and low voltage batteries **Page 159**
- High-voltage battery diagnosis and testing **Page 182**
- Contactors and SMR operation **Page 186**
- Vehicle operating information and CAN Bus **Page 197**
- Motor generator operation **Page 203**
- Inversion and rectification **Page 210**
- DC to DC converter operation and testing **Page 214**
- HVAC operation and testing **Page 215**
- Electric vehicle supply equipment (EVSE) operation and testing **Page 216**

By its very nature, the diagnosis and testing of high-voltage components require the systems to be 'live' for many activities. This significantly increases the risks involved when working on or around high-voltage systems.

The hazards of electricity are well-known, but they can be easily ignored due to its invisible nature. This can lead to complacency unless the fundamental operation of electric vehicles is understood. Even with this understanding, caution is necessary. Do not rely on any safety systems designed for protection; instead, take precautions to minimise the risk of injury or death. Always evaluate the risks associated with any activity and implement measures to eliminate or reduce the hazards involved in any task.

Additional risks associated with working on, or around electric vehicles may include:

- Electrocution
- Strong magnetic fields
- Falling from height
- Injuries caused by incorrect manual handling techniques
- Chemicals
- Gases or fumes
- Hybrid engine systems starting unexpectedly
- The silent movement of electric vehicles while in use

Never attempt to work on a high-voltage electrical system unless you have received adequate training.

High-Voltage Diagnosis and Testing

Information Sources

The diagnosis and testing of high-voltage components on hybrid and electric vehicles necessitate a reliable source of technical information and data. Before starting and during the repair process, it's crucial to have comprehensive information about these systems.

Potential sources of information may include:

Table 5.1 Possible information sources	
Verbal information from the driver	Vehicle identification numbers
Service and repair history	Warranty information
Vehicle handbook	Technical data manuals
Workshop manuals/Wiring diagrams	Safety recall sheets
Manufacturer specific information or data sheets	Information bulletins
Technical helplines	Advice from other technicians/colleagues
Internet	Parts suppliers/catalogues
Jobcards	Diagnostic trouble codes
Oscilloscope waveforms	On vehicle warning labels/stickers
On vehicle displays	Temperature readings

Always compare the outcomes of any high-voltage testing and diagnosis against appropriate data sources. Regardless of the information or data source you choose, it's crucial to assess its usefulness and reliability for your repair work.

If you need to replace any electrical or electronic components, always ensure that the quality meets the original equipment manufacturer (OEM) specifications. If the vehicle is under warranty, using inferior parts or making deliberate modifications might invalidate the warranty. Additionally, fitting parts of inferior quality could affect vehicle performance and safety. You should only replace electrical components if the parts comply with the legal requirements for road use.

Safety Precautions and Procedures When Working on EV/Hybrid Vehicles

Working on or around the high-voltage systems of hybrid and electric vehicles is potentially dangerous. Hazards include, but are not limited to:

- Electric shock
- Burns
- Arc flash
- Arc blast
- Fire
- Explosion

[For details of these hazards, see **Table 3.5**, Chapter 3]

High-Voltage Diagnosis and Testing

Safety Precautions to be Taken Before Carrying Out Any Repair Procedures on Electric Vehicles

The typical voltages used for a range of electrically propelled and hybrid vehicles are 100 to 800V.

Table 5.2 gives examples of recommended high-voltage PPE that should be used when working with electric vehicle drive systems.

Table 5.2 Examples of high-voltage personal protective equipment PPE	
PPE	**Recommendations and use**
High-voltage insulated gloves and over gauntlets	High-voltage PPE gloves are designed to protect the hands from electric shock when working with high-voltage systems. High-voltage PPE gloves are usually made of rubber or synthetic materials that have high dielectric strength, which means they can resist the flow of electric current. ➤ High-voltage PPE gloves are supplied in different class categories based on the maximum voltage they are rated to withstand, ranging from Class 00 (500 volts) to Class 4 (36,000 volts). The minimum class required for working on and around the high-voltage systems of electric vehicles is Class 0 (1,000 volts). ➤ High-voltage PPE gloves should be worn with leather protectors over them to prevent damage from abrasion, cuts, or punctures. However, they provide no extra protection against electric shock, so they should be used in addition to, but not instead of, high-voltage insulated gloves. ➤ High-voltage PPE gloves need to be tested before each use to ensure they are free of defects or holes that could compromise their insulation. The gloves can be inflated at the wrist, and then rolled up to ensure that no air leaks are present. ➤ When a new pair of gloves is commissioned, the date they are brought into use should be recorded. Once exposed to air and UV, the materials begin to degrade. Most glove manufacturers will stipulate that the gloves will require recertification or replacement after 6 months. This can sometimes be extended to twelve months if they are not used frequently and are kept according to the manufacturers' recommendations in their original packaging. ➤ Cotton inner-liner gloves may also be used, which reduce perspiration and increase hygiene. However, they provide no extra protection against electric shock, so they should be used in addition to, but not instead of, high-voltage insulated gloves.
Eye protection	Eye protection is available in various formats. It is primarily designed to provide protection against impact, heat, chemicals, and fumes. The most appropriate form of eye protection should be chosen for the type of task being undertaken. ➤ Safety glasses are a type of protective eyewear designed to protect your eyes from small flying objects, such as splinters or dust, as well as chemicals and light. They usually have side shields or wrap around the temples to prevent objects from entering from the sides. ➤ Safety goggles are a type of protective eyewear designed to protect your eyes from small flying objects, such as splinters or dust, as well as chemicals and light. They fit tightly against the face and form a seal around the eyes, preventing any particles or liquids from entering. ➤ Safety face shields are a type of protective eyewear designed to protect your face from small flying objects, chemicals, light, and heat. They can be worn over glasses or goggles to provide extra protection.

High-Voltage Diagnosis and Testing

Table 5.2 Examples of high-voltage personal protective equipment PPE

Overalls or Workwear	
	Specific workwear or overalls provide an additional layer of protection between the user and potential hazards. Depending on the materials used, this protection could include containment of loose clothing and the covering of areas of bare skin, resistance to chemicals or limited fire protection. It is recommended that overalls or clothing do not use metal fastenings when working on or around the high-voltage systems of hybrid and electric vehicles. Metal fastenings may increase the risk of short circuit, leading to electrocution or fire, as well as interfere with the very strong magnetic fields found within the drive systems of electric vehicles.
	Arc flash/blast overalls are a type of PPE that is designed to protect the wearer from the hazards of arc flash and blast, which are intense bursts of heat, light and sound that can occur when an electrical fault causes an arc between two conductors. Arc flash and blast can cause severe burns, blindness, hearing loss, shock and even death. Arc flash/blast overalls are usually made of flame-resistant fabrics that do not ignite, melt, or drip when exposed to high temperatures. They may also have reflective strips or patches to increase visibility in low-light conditions.
Safety footwear and high-voltage overshoes	Safety footwear is a type of PPE that is designed to protect your feet from injuries caused by impact, penetration, heat, cold, chemicals or electricity. Safety footwear is usually made of leather, rubber or synthetic materials that have different properties and resistance levels against various hazards. Some common features of safety footwear include: ➤ Protective toe caps: These are made of steel, aluminium, composite, or plastic materials that can withstand high forces and prevent crushing or puncturing of the toe area. ➤ Penetration-resistant midsoles: These are made of steel, textile or composite materials that can prevent sharp objects from piercing through the sole of the shoe. ➤ Electrically insulated soles: These are made of non-conductive materials that can inhibit electricity and reduce the possibility of electric shock. They are also often marked with a green triangle symbol to indicate their electrical safety rating.

	High-voltage overshoes are a type of protective footwear designed to protect feet from electric shock when working with high-voltage systems. High-voltage overshoes are usually made of rubber or synthetic materials that have high dielectric strength, which means they can resist the flow of electric current. High-voltage overshoes are worn over regular shoes and cover the entire foot area. Similar to high-voltage insulated gloves, the overshoes have different class categories based on the maximum voltage they can withstand, ranging from Class 00 (500 volts) to Class 4 (36,000 volts). The minimum class required for working on and around the high-voltage systems of electric vehicles is Class 0 (1,000 volts).

[Other safety precautions that should be observed when working with vehicle high-voltage systems are shown in Chapter 4 **Table 4.3**].

High-Voltage Diagnosis and Testing

Specialist Tooling and Equipment

For the list of specialist tooling and equipment that should be used when working on the high-voltage systems of electric vehicles, [see Chapter 4 **Table 4.4**].

High and Low Voltage Batteries

In order to understand the construction and operation of batteries, a description of atoms is useful. Although not necessarily scientifically correct, a simple explanation using the Rutherford Model can assist comprehension of electricity and its actions. It is based on a description proposed in 1911 by the New Zealand-born physicist Ernest Rutherford, where atoms are described as a tiny dense positively charged core called the nucleus, surrounded by light negatively charged particles called electrons. This can be imagined like planets revolving around a sun in a solar system.

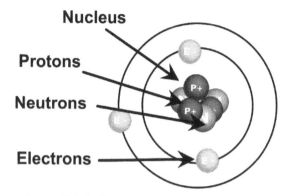

Figure 5.1 A Simple Atom (Rutherford Model)

The number of positively charged particles in its nucleus determines its identity as an element and its position on the periodic table [see Chapter 1]. The number of electrons in orbit around the nucleus gives the element its personality. If the atom contains fewer electrons than protons in its nucleus, it becomes positively charged and it is known as ionised. Like the poles of a magnet, a positively charged atom will attract negatively charged atoms, and in some circumstances, strip electrons from others to fill deficiencies in its own electrons. When electrons move from one atom to the next, this produces electric current.

Some elements or compounds allow the easy transfer of electrons between atoms; these are known as conductors.

Other elements or compounds restrict the movement of electrons between atoms; these are known as insulators.

 Certain elements have properties that function as either conductors or insulators. They can even be switched between these two states, serving as controls for electronic systems. These versatile elements are known as semiconductors.

The principle of a battery often involves a construction that provides a chemical imbalance in the materials it contains, creating positively and negatively charged zones. A compound called an electrolyte (often liquid) that facilitates the transfer of electrons is used as a conduit between the zones, which are then contained in compartments known as cells. If a cell is connected to a circuit, the positive zone will attract electrons from the negative zone and through the external circuit creating electric current. The moving electrons in the circuit (current) create an electromotive force (EMF) or pressure proportional to the elements or compounds used in the cell design, providing a nominal voltage. This voltage or pressure, in association with current flow, can be converted into useful work by a component known as a consumer. The consumer will convert the electrical energy into either heat, magnetism, or a chemical reaction.

Current in a complete electrical circuit will continue to flow until the elements contained within the cell have similar amounts of electrons and chemical consistency; this process is known as discharging. When a cell is fully discharged, it is often referred to as flat. In some battery construction types, the chemical process is reversible, and the battery can be recharged. This is achieved by applying a voltage potential (pressure) greater than that coming from the battery cell, forcing the electrons back to their original zones.

High-Voltage Diagnosis and Testing

ANALOGY

The principles within a battery cell can be likened to an economic state with rich and poor areas of society.

> The rich being the negatively charged zone and the poor being the positively charged zone.
> The electrolyte could be seen as a folk hero, like Robin Hood, who seeks to distribute wealth between the rich and the poor and improve the lives of those less fortunate.

In the image shown, wealth (electrons) are taken from the rich by Robin Hood and given to the poor. The poor spend their newfound wealth in the economy, benefiting the rich but to a lesser extent. When both areas of the economic state are equally wealthy, the process will stop, and this is similar to a battery being flat.

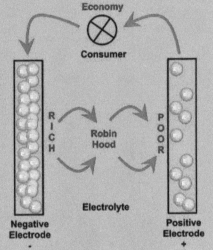

Notes

It is not possible to overcharge a battery cell containing an aqueous (water-based) electrolyte, as excess charge will create a reaction within the liquid, causing it to vaporise or gas. This means that chemical charge will be limited by the constituent elements. This process is known as a redox shuttle.

Pressure and chemical gases must be carefully controlled and managed within the cell design and charging process, as excess amounts could lead to fire if exposed to a source of ignition or even explosion.

SAFETY

The chemicals used in battery electrolyte are often highly toxic and corrosive. Any leakage or spills must be carefully managed and dealt with following prescribed methods, wearing appropriate chemical-resistant personal protective equipment (PPE). Any waste must be correctly disposed of in line with environmental standards and legal requirements. [See Chapter 3].

Battery Capacity

The amount of energy that a battery can hold is often referred to as battery capacity. Normally measured in ampere-hour or amp hour (Ah), it is an estimation of the electrical current that a battery can deliver for a set period of time. This quantity is one indicator of the total amount of charge that a battery can store and deliver at its rated voltage. The ampere-hour (Ah) value is the product of the discharge current (in amperes) and the time (in hours) for which this discharge current can be sustained by the battery. For example, a vehicle battery rated as 100 Ah should contain enough electricity to provide:

> 100 amps for one hour.
> 1 amp for 100 hours.
> 10 amps for 10 hours.
> Any other combination that multiplies together to make 100 (e.g. 25 amps for 4 hours).

Small battery cells may have their capacity expressed as milliampere-hour or milliamp hour (mAh).

High-Voltage Diagnosis and Testing

> The expression of capacity in amp hours could be misleading, as its value is only linear when operating at its rated voltage value. Once the voltage drops below a set limit, the discharge values no longer apply, and current flow will normally increase depending on demand.
>
> Capacity using the amp hour rating is difficult to compare between vehicle types, as voltage and power will vary between vehicle style and use. Although not perfect, power capacity, measured in watt-hour (Wh), provides a more general/comparative value.
>
> Non-rechargeable batteries hold more energy than rechargeable ones but cannot deliver high load currents.

Watt-hour (Wh)

Power capacity is expressed in watt-hour (Wh). It is better at showing a comparison of battery capacity between different vehicle styles and types, regardless of the voltage used in the design of an electric vehicle. It is calculated by multiplying voltage by amp hour (Ah).

V x Ah = Wh

Specific energy - the capacity a battery can hold, is expressed as watt-hour per kilogram (Wh/kg).
Specific power - the ability to deliver power, expressed as watt per kilogram (W/kg).

Battery Chemistry

Battery chemistries and uses will vary from application to application, however, three main styles are in common use with electric vehicles:

Lead acid (low-voltage) auxiliary systems.

This is one of the oldest rechargeable battery designs. They have a very rugged design and are forgiving when worked hard. They are comparatively cheap and operate over a large temperature range. Unfortunately, they have a low specific energy for weight and volume, making them unsuitable for drive batteries in most electric vehicles. They have a relatively limited life cycle and will produce gases during the process of charging, so they will require a method of ventilation within their design.
The nominal cell voltage of a single lead acid cell is approximately 2.1 volts.

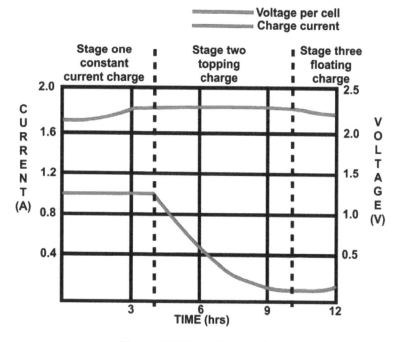

Figure 5.2 Charging Lead Acid

High-Voltage Diagnosis and Testing

Nickel metal hydride (NiMH) high-voltage, used mainly with hybrid drive.

Nickel metal hydride (NiMH) batteries have a much higher specific energy than lead acid batteries, however, they are still heavy relative to the amount of power available. As a result, this chemistry type was mainly restricted to hybrid electric powertrains, where it is used to only supplement an alternative power source such as an internal combustion engine. The chemistry is mildly toxic, highly corrosive, and can be difficult to charge.

The nominal cell voltage of a single NiMH cell is approximately 1.25 volts.

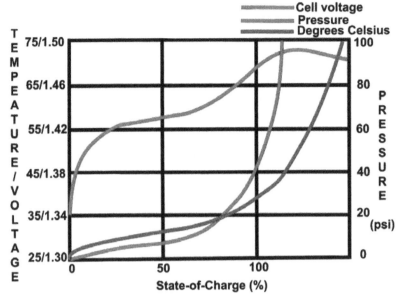

Figure 5.3 Charging NiMH

Lithium ion (Li-ion) high-voltage, used mainly with electric vehicle powertrains.

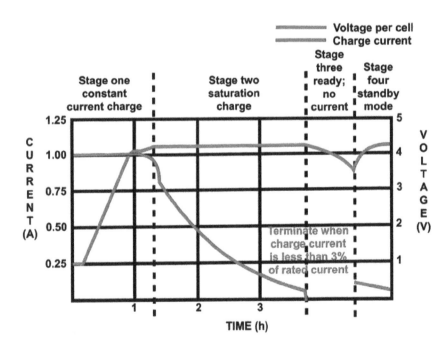

Figure 5.4 Charging Li-ion

Lithium ion (Li-ion) battery cells are constructed from the lightweight metallic element, lithium. Lithium is a highly reactive element, and its volatile nature makes it a very useful component for exchanging energy within a battery cell. Its power capacity for size, weight, and volume makes it an ideal drive battery for electric vehicle powertrains. Its charge and discharge rates need to be carefully controlled within set limits, to avoid cell damage or potential thermal runaway. Depending on the exact chemistry, a single lithium-ion cell has a **nominal voltage** of approximately 3.6 to 3.75 volts.

Degradation

Regardless of battery design or chemistry, the process of charging and discharging will cause a battery cell to age and degrade. This will affect its ability to store and distribute electrical energy. When a battery is assembled new, cells are chosen in matched sets to produce a stable and consistent pack. However, cells will degrade at different rates when in use, and this relates to its state of health (SoH).

Different battery chemistries will degrade at different rates.

High-Voltage Diagnosis and Testing

An analogy that can be used to explain the degradation of a battery cell is to liken it to the use of an electric kettle. The kettle uses **resistance** in an element to convert electrical energy into heat, which can be used to boil water.

When an electric kettle is brand new, it will be the most efficient and effective as it will ever be at boiling water. Boil the kettle several hundred or thousand times and a scale from minerals in the water will begin to form around the element and on the inside walls of the kettle. Over time, this scale will reduce the efficiency of the heating element and the overall capacity of the kettle itself.

Although not scale, the breakdown of chemicals and elements works in a similar manner, affecting the efficiency and capacity of a battery cell.

Solid State Battery Cells

A solid state battery cell is one that uses a solid electrolyte rather than the liquid or **polymer gel** found in other lithium ion types. A battery **electrolyte** is a conductive chemical mixture that helps the flow of electrons between the electrodes of a battery cell. The positive electrode of a cell is known as the **cathode**, and the negative electrode is known as the **anode**. The electrolyte sits between the anode and cathode and will normally contain a porous electrical insulator known as a separator, which stops the electrodes from touching each other, helping to prevent short circuits. A solid electrolyte alters some of the battery's characteristics when compared to liquid or gel. It will often have a higher energy capacity, quicker charge time, longer lifespan, and potentially be safer than a liquid or gel equivalent.

The words anode and cathode are derived from Greek words that mean "way up" and "way down" respectively. They were coined by William Whewell, an English **polymath**, in 1834 for Michael Faraday, the English chemist and physicist, who used them to describe the direction of the electric current in a circuit.

Nominal voltage - a value assigned to a circuit or system for the purpose of designating its voltage class. It is not the precise operating or rated voltage but rather an "approximate" or "average" level around which the actual voltage can vary while still allowing satisfactory operation of equipment.

Resistance - a measure of the opposition that a substance offers to the flow of electric current.

Polymer gel - a solid composed of at least two components, one of which is a polymer that forms a three-dimensional network through covalent or noncovalent bonding in the medium of the other component, which is typically a liquid with elastic properties.

Cathode - the positive terminal of a battery electrode.

Anode - the negative terminal of a battery electrode.

Polymath - a person who possesses a wide range of knowledge or learning across multiple subject areas.

High-Voltage Diagnosis and Testing

Types of Li-ion Batteries

The chemistry in a lithium ion (Li-ion) battery can vary between manufacturers and can be tailored for its properties and use. For example:
- Specific energy
- Specific power
- Safety
- Performance
- Life span
- Cost

The cell will consist of a metal oxide cathode, a graphite anode, and a non-aqueous electrolyte with separator. The metal oxide will vary depending on the manufacturer and the potential use, and some examples are shown in **Table 5.3**. The electrolyte solution consists of two or more chemicals combined to make a compound; one is a solute, which is dissolved in a solvent. Solutions can be gaseous, liquid, or solid. The most common type of electrolyte in a Li-ion battery cell consists of lithium hexafluorophosphate as the solute, dissolved in a solvent of either ethylene carbonate or dimethyl carbonate, which are both liquid at room temperature.

Unlike the chemical reactions in most battery cells, the reactive properties of lithium and its tendency to give up electrons and become ionised are capitalised upon. When situated within a metal oxide, lithium is in a stable state with all its electrons in orbit. This could be considered its 'happy place'. When an external potential (charge) is applied, lithium is directed to a storage area at the anode, made from graphite. The electrolyte solution promotes the movement between the cathode and anode; however, it will only allow ionised lithium through. This means the free electrons need to take an alternative path to the anode through an external circuit. When charged, the additional electrons rejoin their lithium ions (returning to complete de-ionised lithium) in the graphite storage area. The graphite storage area is considered the 'unhappy place' for the lithium, and it would like to return to the metal oxide at the first opportunity.'

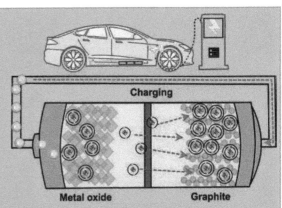

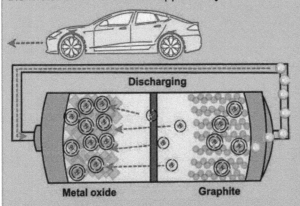

When connected to an external load/consumer (a motor for example), the lithium is able to travel back to its 'happy place' in the metal oxide cathode, however, once again the electrolyte will only allow ionised lithium through, meaning that the spare electrons need to take an alternative route through the external circuit, powering the consumer with its current flow.

The actions inside a Li-ion cell are like having two chambers connected with a pipe and valve. When charged, one chamber is pressurised, having a higher pressure than the other. As soon as the valve is opened, pressure moves from the high chamber to the low chamber via the external pipe.

High-Voltage Diagnosis and Testing

Lithium Battery Types and Chemistry

Table 5.3 shows some typical lithium battery chemistries used in vehicle propulsion and their attributes.

Table 5.3 Battery chemistry

Cathode Material	Abbreviation (see key)	Typical nominal voltage potential	Approximate energy density Wh/Kg & Wh/l	C-rate (charge rate)	Typical ambient temperature range	Approximate thermal runaway onset	Key*
Lithium nickel manganese cobalt oxide	NMC*	3.6 volts	150-220 Wh/Kg 260-400 Wh/l	3C	-30°C to 60°C	210°C	What do the letters in the cell variants stand for: L = Lithium C = Cobalt O = Oxide (O$_2$) N = Nickel A = Aluminium F = Iron P = Phosphate M = Manganese T = Titanate
Lithium nickel cobalt aluminium oxide	NCA*	3.6 volts	200-260 Wh/Kg 490-670 Wh/l	2-3C	-20°C to 60°C	150°C	
Lithium iron phosphate	LFP*	3.2 volts	90-120 Wh/Kg 100-170 Wh/l	30C	-30°C to 60°C	270°C	
Lithium manganese oxide	LMO*	3.7 volts	100-150 Wh/Kg 240-360 Wh/l	3-10C	-20°C to 60°C	250°C	
Lithium titanate	LTO*	3.7 volts	50-80 Wh/Kg 170-230 Wh/l	10C	-30°C to 75°C	Less susceptible to thermal runaway	

The C-rate is the unit used to measure the speed at which a battery is fully charged or discharged. For example, charging at a C-rate of 1C means that the battery is charged from 0% to 100% in one hour.

Sometimes the cell variant abbreviation is followed by a set of numbers indicating their relative proportions. For example, NMC111 means that the metals have equal amounts, while NMC532 (or NCM523) means that nickel has 50%, manganese has 30%, and cobalt has 20% of the total metal content.

Figure 5.5 A High-Voltage Drive Battery

High-Voltage Diagnosis and Testing

Cell Shape Design

Battery cells are often packaged in three main shapes:

Pouch

In this design, the battery cell is contained in a form of plastic pouch. This can save on materials, weight, and reduce cost in some circumstances. The chemical reactions that take place during charging and discharging create heat, which in turn will create pressure.

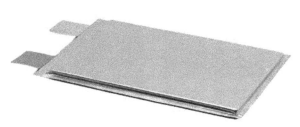

The expansion of the materials inside the cell will make the pouch swell, so when combined into modules, an allowance for this expansion must be a design consideration. As the cells cool, the internal pressures will reduce, and the pouches will return to their original shape. Careful monitoring of charge, discharge, and heat will be a safety factor in the use of this design, to ensure that any expansion of the pouch does not lead to the cell bursting.

Figure 5.6 A Pouch Style Battery Cell

Prismatic

In this design, the battery cell is contained in a rectangular block. The word prismatic originates from the Greek word for 'sawn' and relates to prisms.

The design and shape of a prismatic cell is often considered optimal for the layout of battery modules or packs. This is because they can be assembled in a very compact arrangement, with little wasted space in between. However, this can often lead to difficulties with weight and cooling. Another issue with prismatic cell design is that any internal pressures created by heat during the process of charging and discharging can cause the cell casing to swell. As a result, prismatic cells or modules are often assembled so that they sit within a compressed or clamped format. Prismatic cells should never be charged or discharged in an uncompressed state.

Figure 5.7 A Prismatic Style Battery Cell

Cylindrical

In this design, the battery cell is contained within a cylinder.

The cylindrical shape is naturally resistant to internal pressures; therefore, chemical reaction and heat are less likely to make the cell swell. However, due to its shape, they tend to take up more space inside a battery module or pack than an equivalent prismatic cell. Cylindrical cells are often much smaller than prismatic or pouch styles when used within the battery design of an electric vehicle, meaning that more individual cells are used to make up the pack, connected in parallel to provide the capacity required. The designation of cylindrical type cells is often based on their size and shape, for example, an 18650 is 18 millimetres in diameter and 65 millimetres in length.

Figure 5.8 A Cylindrical Style Battery Cell

High-Voltage Diagnosis and Testing

Layouts

Depending on the requirements of the vehicle, battery packs are assembled so that cells are connected in series to supply the voltage demand. The physical size of the cells connected in series will determine the battery capacity; the larger the cell, the more electricity it can hold.
Some batteries are manufactured with relatively small cells. The series construction provides the required voltage, but many other cells are connected in parallel to provide the capacity.

> In battery assembly, cells connected in series increase voltage, while parallel connections boost capacity without raising voltage.
> The optimal assembly combines series and parallel connections for the right voltage and increased capacity.
> Parallel connections also mitigate performance issues caused by individual cell depletion, as they distribute current flow across cells, preventing imbalance and excessive temperature hotspots.
> In contrast, series-only assemblies can be affected by individual cell depletion, leading to restricted current flow and potential temperature hotspots.

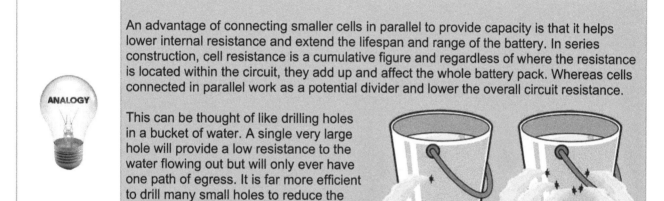

An advantage of connecting smaller cells in parallel to provide capacity is that it helps lower internal resistance and extend the lifespan and range of the battery. In series construction, cell resistance is a cumulative figure and regardless of where the resistance is located within the circuit, they add up and affect the whole battery pack. Whereas cells connected in parallel work as a potential divider and lower the overall circuit resistance.

This can be thought of like drilling holes in a bucket of water. A single very large hole will provide a low resistance to the water flowing out but will only ever have one path of egress. It is far more efficient to drill many small holes to reduce the resistance to flow and provide more exits for the water to leave.

When combined series/parallel batteries are assembled, two main layouts can be considered:

Parallel then series - the most common design used for electric vehicles. In this design, the parallel cells are first grouped together, and then connected in series to form the battery pack. Balancing currents move between the cells connected in parallel and a single battery management unit (BMU) can be used for the whole pack.

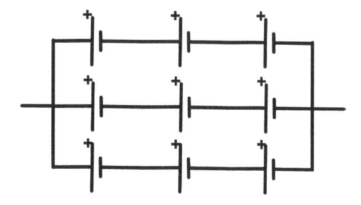

Figure 5.9 Parallel Then Series Construction

High-Voltage Diagnosis and Testing

Series then parallel - this assembly gives more flexibility in design for large battery packs. The cells are first connected together in series, then the series groups are joined in parallel. Current flows between the series-linked cells, meaning a more complex measurement of voltage is required in order to maintain the battery pack. A separate battery management unit (BMU) is required for each set of linked series cells, and in turn, these BMUs need a master controller to manage them.

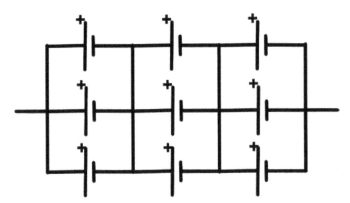

Figure 5.10 Series Then Parallel Construction

Some advantages and disadvantages of series and parallel construction are described in **Table 5.4**.

Table 5.4 Advantages and disadvantages of series and parallel construction			
Series advantages	**Series disadvantages**	**Parallel advantages**	**Parallel disadvantages**
☑ Series construction increases the voltage and boosts the power delivered rather than the current.	☒ The failure of a single cell will stop the entire battery from working.	☑ Wiring in parallel increases battery capacity and vehicle range.	☒ The size of the battery and its construction may increase, adding additional weight and reducing performance and efficiency.
☑ Increasing the voltage will mathematically increase the vehicle range if the power remains the same.	☒ Internal resistance becomes a cumulative/compound figure.	☑ Wiring in parallel reduces internal resistance and improves current flow and efficiency.	☒ The need for more physical resources, chemicals and components increases the cost of production and the environmental impact.
☑ Increasing the voltage will allow the construction of more powerful systems and keep the range similar.	☒ Increased resistance due to additional wire reduces operating efficiency.	☑ The total voltage doesn't change, which simplifies battery management and reduces the need for higher-voltage-rated components.	☒ More components to recycle at the end of life adds to the complexity and cost of disposal.

High-Voltage Diagnosis and Testing

Table 5.4 Advantages and disadvantages of series and parallel construction			
Series advantages	**Series disadvantages**	**Parallel advantages**	**Parallel disadvantages**
☑ Series construction enables efficient packaging within the vehicle's body to maximise capacity.	☒ Increased heat generation in the system can cause thermal runaway or damage.	☑ Parallel construction enables efficient packaging within the vehicle's body to maximise capacity.	☒ The need for additional monitoring and balancing of parallel cells increases the demand for battery management systems and controllers.
☑ Series construction offers flexibility in where the battery is placed within the vehicle body and therefore the effect that has on power efficiency and vehicle handling characteristics.	☒ The necessity for higher-voltage-rated components in the entire system increases cost and complexity.	☑ The battery's placement within a vehicle can affect power efficiency and vehicle handling characteristics, and parallel construction offers more flexibility.	☒ More cells may increase the risk of failure, imbalance, and thermal runaway, which can compromise safety and reliability.
☑ Higher voltages allow more power to be transferred with less loss over the same diameter and mass of electrical cable.	☒ Series construction prevents the ability to use DC fast-charging stations of a lower voltage without incorporating some type of DC-DC boost converter in the on-board charger.	☑ If one of the parallel cells fails to operate, the remaining cells can still provide power, which enhances reliability and safety.	☒ Packaging may lead to increased heat generation and dissipation issues, which can affect the lifespan and performance of the battery.
☑ Series construction results in less capacity, meaning that charging is quicker.	☒ In a series-connected battery system, a converter is needed to achieve low voltages for auxiliary devices.		
☑ Higher voltage means less cross-sectional area in copper wiring will be needed to handle the same amount of power.			

High-Voltage Diagnosis and Testing

State of Charge (SoC)

A key component of the amount of energy in a battery cell and therefore the power it is able to supply is its state of charge, often abbreviated to SoC. This is calculated as a percentage from a measured voltage, compared with the cell's possible chemically available original voltage. The battery management system (BMS) constantly monitors the voltage from the cells, normally in batches known as modules or blocks, so it can give an estimate of the remaining energy to the driver in a similar manner to a fuel gauge on an internal combustion engine (ICE) propelled vehicle. The state of charge is also used to maintain the balance of battery cells or modules. This is important because as the battery is charged and discharged, the cells will not empty and fill at the same rate. If left unbalanced, over a period of time, the battery's cell state of charge will become uneven. The chemistry of the high-voltage battery cells means they are susceptible to over and under charge, which may result in irreversible damage and failure. Unless the SoC is kept within an operating voltage window, the usefulness will degrade quickly. [See battery balancing for more details].

State of charge (SoC) is often confused with a battery cell being either full or flat. However, SoC is actually an operating window where the voltage sits between an upper and lower value. These values are dictated by the cell chemistry or system operating requirements.

Below is a table summarising the states of charge (SoC) and corresponding voltages for different battery chemistry types used in electric vehicles:

Battery Type	100% SoC Voltage	50% SoC Voltage	0% SoC Voltage
Lead Acid	12.6V	12.3V	12.0V
Nickel Metal Hydride (NiMH)	1.4V	1.2V	1.0V
Lithium Ion	4.2V	3.6V	3.0V

Please note that these voltages are approximations and can vary based on factors such as temperature, discharge rate, and specific battery design. Always refer to the manufacturer's specifications for the most accurate information.

State of health (SoH)

State of health (SoH) is an indication of the battery's usable lifespan as an energy source for the propulsion of an electric vehicle. As the battery ages, charging and discharging causes the internal resistance of individual cells to increase. The natural internal resistance caused by the cycle of charge and discharge reduces the battery's capacity and effectiveness with age. The driver may experience the ageing of the battery as a loss of usable driving range. In fact, when the SoH reaches approximately 70%, the battery may need to be replaced to maintain an acceptable driving distance range.

Cell ageing and its effects are similar to giving an energy drink to both a teenager and a 90-year-old.

Although they will have both received the same amount of energy, it is likely that the teenager will be able to make more effective use of the energy provided.

High-Voltage Diagnosis and Testing

Another issue caused by cell internal resistance is voltage drop.

According to Ohm's law, the current flow demanded by the components, such as drive motors, multiplied by the resistance, causes the voltage value of the battery cells to drop. When the battery is new and the resistance is low, the voltage drop is kept within acceptable limits. However, as the battery ages and the internal resistance increases, so does the voltage drop. If the demand on a battery cell causes the voltage to fall below a certain level, it may be irreversibly damaged. If any cells within the battery pack connected in series fail, this will immediately prevent current flow, stopping the battery and vehicle operation.

> When a vehicle's high-voltage drive battery has fallen below its useful state of health (SoH) and drive range has significantly reduced, it may still be suitable for repurposing. This means the battery may not need disposal or recycling but could instead be repurposed as an electrical storage medium, like solar energy storage in a power-wall, for example.

Battery Management Systems (BMS)

The battery management system (BMS) is an electronic control unit with the main purpose of preserving the performance and lifespan of the high-voltage battery. Without some form of management, the degradation of the battery would be such that its state of health would decline in a relatively short period of time.
Its main functions are to monitor state of charge and temperature and to provide output control over thermal management, charge/discharge in conjunction with the EV control unit and cell balance.

Figure 5.11 A BMS

Thermal management normally involves cooling the high-voltage battery, as in use, a byproduct of the charge/discharge process is heat. However, it can also preheat batteries in colder climates. Thermal management will often make use of air flow, liquid coolant, or air conditioning to maintain heat within a range specified by the manufacturer, or electric heating elements for pre-conditioning of the modules.

Balancing regulates the state of charge. Depending on manufacturer preference, balancing may be either active or passive.

Battery Data

In order to manage the high-voltage battery correctly and effectively, certain data is required:

Voltage - The overall battery voltage is measured and used to help determine and maintain the state of charge of the battery pack. In a lithium-ion battery, voltage is also measured at the series cell level. This is important as a difference in state of charge between cells or modules connected in series will lead to an imbalance within the pack, and unless correctly managed, will result in both degradation and loss of mileage range. [See battery balance].

> Due to chemistry, Nickel Metal Hydride (NiMH) cells will naturally balance within reason; [see **Redox Shuttle**].
> This means that the measurement of individual cell voltage is less critical to the overall management of the battery pack. Therefore, voltage may only be measured at either module level, or grouped modules often referred to as blocks.

High-Voltage Diagnosis and Testing

Current - The current leaving or returning to the battery is often monitored in one of two ways.

The most common measuring device is an inductive sensor. This takes advantage of the **Hall Effect**, [see Chapter 1]. In a similar manner to an amp clamp, the **inductive sensor** is mounted around a main battery cable or bus bar at the battery unit itself. (Current is the same regardless of whether you measure it on the positive or negative side of the circuit.) The magnetic field created as current flows through the cable or **bus bar** creates a voltage differential in the Hall Effect sensor, which is proportional to the magnetic field strength created by current flow. This voltage differential can then be used as an estimation of current flow.

A **shunt** is another method of measuring current flow entering or leaving the high-voltage battery. A shunt is connected in series to either the positive or negative high-voltage battery connection and may resemble a bus bar or large fuse with attached electronics. It measures the voltage drop across a known resistance value, and then uses Ohm's law to calculate current flow: voltage divided by resistance equals current.

An inductive sensor is not as accurate as a shunt; however, it is far cheaper to produce and therefore the most common form of current sensor used.

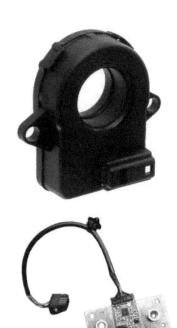

Figure 5.12 Current Sensors

Temperature - A byproduct of the charging and discharging process is heat. Temperature must be carefully monitored by the battery management system (BMS) to ensure that the high-voltage battery cells remain in the optimum temperature range. Temperature can also be an indication of battery load. Often measured at several different positions within the high-voltage battery, **negative temperature coefficient (NTC)** thermistors provide feedback to the BMS and are compared with an **ambient temperature** measurement to accurately assess heat energy. If temperatures exceed preset limits, the BMS will implement thermal management to help regulate battery conditions.

Insulation and Isolation - A vital safety feature of the high-voltage drive system is the effective monitoring of **insulation** and **isolation**.

This ensures, as far as possible, that if there were to be a breach of insulation in the high-voltage system to the vehicle frame or chassis, the driver can be informed and, if necessary, shut down to protect users and operators. It often consists of a unit which measures the voltage potential difference between the high-voltage and low-voltage systems and chassis. In general terms, when operating, there should be a large potential voltage difference between both systems if they are adequately isolated from each other.

Using current flow measurement from the battery, this voltage potential can be used to calculate and display an insulation resistance reading in diagnostic live data.

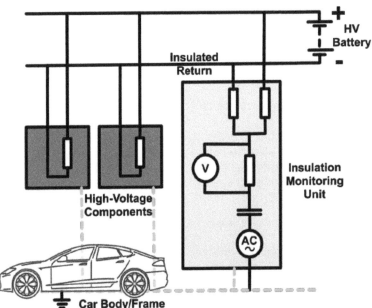

Figure 5.13 Insulation Measurements

High-Voltage Diagnosis and Testing

 Although insulation and isolation measurement are constantly monitored, as with any safety system, it has the potential to fail or provide false results. Always assume that safety systems have failed, and take all required precautions, including the use of high-voltage PPE, until they have been properly tested and assessed for operation and accuracy.

SoH Calculations and Approximations

The Battery Management System (BMS) calculates the State of Health (SoH) of an electric vehicle's battery by monitoring and controlling the battery's state. State of health is an estimation of the battery's overall degradation due to use and the aging process.

Redox Shuttle - short for reduction-oxidation reaction, is a type of chemical reaction in which the oxidation states of atoms are changed, often resulting in the breakdown of electrolytes into gaseous states.

Hall Effect - the production of a potential difference, known as the Hall voltage, across an electrical conductor. This occurs when there is an electric current in the conductor and a magnetic field applied perpendicular to the current. Discovered by Edwin Hall in 1879.

Inductive sensor - a device that uses the principle of electromagnetic induction to detect or measure objects. It operates by generating an oscillating electromagnetic field created by a moving magnetic object, such as a metal target.

Bus bar - an electrical conductor, or a group of conductors, that serves as a central hub in power distribution. It is typically a metallic strip or bar, often made of copper or aluminium, and is used to consolidate electric power from incoming feeds and distribute it to outgoing feeds.

Shunt - also known as a shunt resistor, is a device that provides a low-resistance path for electrical current to flow through in a circuit. It is typically used to measure the amount of current flowing by creating a voltage drop that can be easily measured and correlated to the current.

Negative Temperature Coefficient (NTC) - a characteristic of materials where their electrical resistance decreases as their temperature increases.

Ambient temperature - the temperature of the air surrounding an object or within a particular area.

Insulation - a method that resists the flow of electric current. It is used to isolate electrical conductors from each other and from any other conductive materials or grounded surfaces.

Isolation - the process of separating a high-voltage electrical circuit from the rest of the vehicle. This separation is done to protect users from high-voltages and to prevent damage to electrical equipment. It ensures that no direct current (DC) or unwanted alternating current (AC) can transfer between the isolated parts of the vehicle.

High-Voltage Diagnosis and Testing

Lithium-ion battery cells will degrade through use and as part of the natural aging process that occurs through charge and discharge.

Issues include:

- **Solid electrolyte interphase (SEI) also known as solid electrolyte interface** - a layer that forms on the surface of the negative electrode. It can protect the electrode from further degradation, but it will increase internal resistance.
- **Lithium plating** - the loss of metallic lithium on the negative electrode during charging, especially at low temperatures, high currents, or high states of charge.
- **Electrode cracking** - the fracture of the active materials or the binder on the electrode. This can be caused by mechanical stress, temperature, or volume change.
- **Electrode dissolution** - the loss of active materials or additives from the electrode into the electrolyte. This can be caused by high temperatures, high voltage, or chemical instability.
- **Electrolyte decomposition** - the breakdown of the electrolyte into simpler or different compounds. This can be caused by high voltages, temperatures, or reactions with the electrodes.
- **Particle cracking** - a type of electrode cracking that occurs at the level of the active materials on the electrode. Particle cracking reduces the contact area between the active material and the current collector, leading to poor performance and reduced capacity.

SEI Solid Electrolyte Interphase (Interface)

Initial Formation → Evolution

Mostly Inorganic / Mostly Organic

Legend:
- Lithium Ethylene Dicarbonate
- Inorganic Carbonates
- Lithium Ethylene Monocarbonate
- Oxalates
- Dilithium Ethylene Monocarbonate
- Ethylene
- Carbon monoxide
- Hydrogen

The state of health (SoH) of a battery can be calculated using different methods. The more battery data that is available, the more accurate the SoH estimation is.

Examples include:

- Counting the remaining number of charge/discharge cycles.
- Coulomb counting (which measures the current that flows in and out of the battery over time).
- Measurements of physical quantities, such as internal resistance, temperature, and voltage.
- Power in versus power out (comparing the energy input and output of the battery).

High-Voltage Diagnosis and Testing

Power out versus power in is a method that can be used to calculate the state of health (SoH) of an electric vehicle battery. It works by comparing the amount of energy that goes into the battery during charging and the amount of energy that comes out during discharging. A degraded battery will have a lower efficiency, meaning that it will lose more energy during charging and discharging due to internal resistance, heat generation, and chemical reactions.

One possible example of a formula using this method is:

SoH = (Energy out / Energy in) x 100%

$$SoH = \frac{E_{out}}{E_{in}} \times 100\%$$

Example: Starting voltage = 200v   Finishing voltage = 199.9v

Power out = 100 Watts ↓   Power in = 100 Watts ↑   **SoH reduced by 0.1%**

Finishing voltage = 100v   Finishing voltage = 100v

Scanning and Live Data

Electric vehicle battery and drive system diagnostic codes and live data should be available through the vehicle's onboard diagnostic system (OBD). To access this information, a suitable diagnostic scan tool will be needed. In order to accurately assess the information provided by the ODB system, a full scan of all available systems needs to be conducted to ensure no current or preexisting diagnostic codes are present within the network. The operation of diagnostic scan tools varies from manufacturer to manufacturer, so always follow any specific instructions or procedures provided.

Although initially designed to provide generic information on emission-related faults with internal combustion engine (ICE) vehicles, **EOBD** or OBD II has been expanded to cover some information relating to hybrid and electric vehicle drive systems. It often operates in its own parallel sub-system, meaning it might contain different information from that provided by the original equipment manufacturer (OEM) data. It is worthwhile scanning for diagnostic trouble codes and live data using both OEM and EOBD/OBD II, because if the vehicle is working in a limited operating strategy (limp) mode, OEM data might show substituted values, whereas EOBD/OBD II will be actual values.

EOBD - stands for European On-Board Diagnostics. It is the European equivalent to the American OBD-II standard.

High-Voltage Diagnosis and Testing

A systematic method for using a diagnostic scan tool is listed below:

1. Place the vehicle into a state of ready mode and check whether any malfunction indicator lamps (MIL) are illuminated on the driver display.

2. Locate the data link connector DLC (which is often inside the passenger compartment within reach of the driver, usually between the centre line of the vehicle and the driver's seat).

3. Plug in the scan tool and were possible set up using the original equipment manufacturer OEM vehicle data (If this is not possible then the generic EOBD/OBD II setting might be used, but this may not give a full list of diagnostic trouble codes or live data).

4. Check the readiness monitors (this will give you an indication that all the required systems have met the prerequisite conditions for testing. If not the vehicle may need be road tested if possible until the required conditions have been met).

5. Following manufacturer's instructions and procedures, conduct a full scan of all available systems.

6. Read and record all/any diagnostic trouble codes (permanent and pending). If any codes indicate a high-voltage safety issue, these should be fully investigated and resolved before proceeding.

7. Read and record any freeze-frame information if available (this will give an indication of the vehicle operating conditions at the time the last fault code was stored and may help you recreate the situation during any simulated operation).

8. Access the live data and compare available values against manufacturers parameters.

9. Clear all diagnostic trouble codes.

10. Operate the vehicle, as long as it is safe to do so (try to recreate any conditions which may have been indicated by available freeze-frame data).

11. Re-scan for trouble codes and concentrate your diagnosis on any codes that have returned (remember to recheck the freeze-frame data).

12. Use further diagnostic testing to check the system and component indicated by any diagnostic trouble code (do not guess or conduct diagnosis by substitution).

13. Find the root cause of the fault and conduct repairs that ensure, as far as is reasonably practicable, the fault will not occur again.

14. Clear any trouble code and re-scan to ensure that no trouble codes have returned (permanent or pending); both before and after road test.

High-Voltage Diagnosis and Testing

Battery Balance

Battery balance is vital to maintaining state of charge and prolonging the state of health of battery cells. The chemistry used in drive batteries is susceptible to over and under-voltage, which can cause irreversible cell damage or even thermal runaway. This means that during the process of charge and discharge, cell voltage needs to be carefully monitored and controlled.

> During discharge, the first cell or module in the battery pack to reach its lower limit will stop the discharge process.
> During charging, the first cell or module in the battery pack to reach its upper limit will stop the charging process.

Due to internal resistance, each battery cell will charge and discharge at a slightly different rate. High internal resistance causes a higher voltage drop when current flows.
This means that discharge is dictated by the weakest cell with the highest internal resistance, and charge is dictated by the cell with the lowest capacity.
Without monitoring and control, the usable amount of energy available for propulsion would fall on a daily basis, artificially reducing driving range. By controlling cell balance, the battery management system (BMS) is able to extend the overall apparent capacity and maintain driving range within limits. Two methods of battery balance are possible: active and passive.

 Overcharging can speed up degradation and potentially cause thermal runaway. For example, with a lithium-ion cell, a slight increase in charging voltage from 4.2V to 4.25V can degrade the battery by approximately 30% quicker.

Passive Battery Balancing

Passive balancing involves removing excess energy from cells with higher voltage.

Once the vehicle and battery have met certain conditions, the voltage of cells or a combination of modules is measured. The cells or modules with higher voltages are then discharged through resistors, removing the excess voltage by converting it to heat. This process brings all the voltages down to the same level, so that during the following charge period, the cells can fill up in a relatively even manner. The advantages of passive balancing include a simple design and architecture, which have considerable cost savings for the design and production of electric vehicles.

The disadvantages of passive balancing are that it is wasteful, essentially throwing away usable energy, and that it balances the battery to the value of the weakest cell.

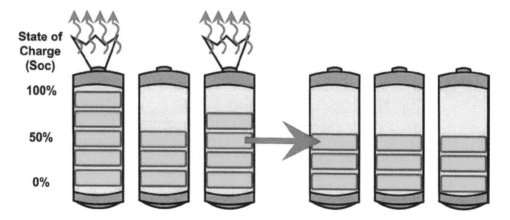

Figure 5.14 Passive Balancing

High-Voltage Diagnosis and Testing

Active Battery Balancing

Active battery balancing involves sharing energy between the cells.

There are several potential designs used in active cell balancing, however, each is more complicated than passive methods, meaning that they tend to be more complex and therefore more expensive. In a similar manner to passive balancing, once the vehicle and battery have met certain conditions, the voltage of the cells or a combination of modules is measured. The cells or modules with higher voltages can be directed, using electronic circuitry, to allow their higher potential to transfer into the cells or modules with a lower state of charge. This process brings all the voltages to the same level, so that during the following charge or discharge period, the cells can fill up or empty in a relatively even manner.

Unlike passive balancing, this method is less wasteful as instead of throwing away energy, it shares the energy out between the cells and maintains voltage at the average of the entire battery pack.

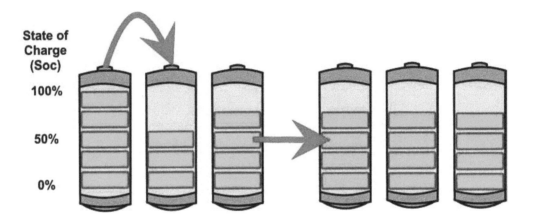

Figure 5.15 Active Balancing

A further method of active balancing can be employed during the charging period of cells. Once a cell has reached its maximum voltage value, electronic circuitry can allow any further charge to bypass that cell or module and continue charging the next.

This process can be repeated, until all available cells have been charged to the same value.

The advantage of active cell balancing is that it brings state of charge to the average of the entire battery pack and is not disposing of any already stored energy within reason.

The main disadvantage is that the complex architecture and electronic circuitry are more expensive to implement.

Charge limiting (cell protection)

Charge limiting is a method that prioritises battery safety over lifespan and involves halting charging or discharging when any cell reaches full charge or minimum voltage respectively. However, this method can significantly reduce range, making it generally unsuitable for electric vehicle battery management.

High-Voltage Diagnosis and Testing

Charge Shunting

Charge shunting is a passive balancing method that discharges a cell's excess voltage as heat through a resistor. The process is activated by a transistor when the cell's voltage exceeds a certain limit.

The discharge of energy aligns it with the lowest cell voltage in the pack, but this process is slow and inefficient due to heat generation and current flow limitations.

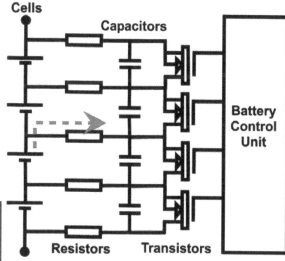

Figure 5.16 Charge Shunting

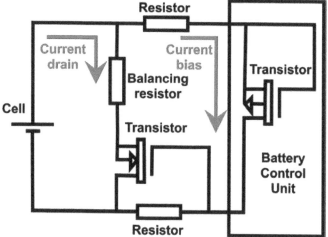

Figure 5.17 Charge Shunting with External Discharge Circuit

Charge shunting speed can be increased by using an external discharge circuit that is separate from the monitoring and control module, but this adds complexity and cost.

Charge Shuttles

The 'charge shuttle' is an active balancing method that uses capacitors to move energy from a higher voltage cell to a lower voltage cell, enhancing the balance of cell voltages. A switched circuit first connects the higher voltage cell to a capacitor, which is charged. The capacitor is then switched to the lower voltage cell and charge is transferred.

This is a complex design and often only allows the transfer of charge between adjacent cells.

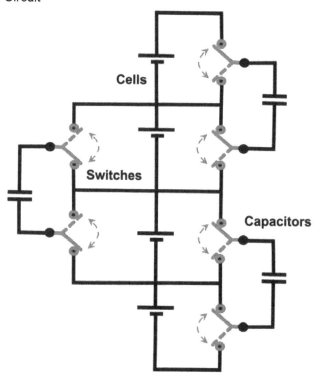

Figure 5.18 Charge Shuttles

High-Voltage Diagnosis and Testing

Inductive Converter

Active cell balancing is often achieved using inductors and or a buck-boost converter. The process involves directing charge from a higher voltage cell to a lower voltage cell. The higher voltage cell is first used to send current to an inductor. Current flow is blocked to the lower voltage cell by a diode. When the circuit to the inductor is switched off, the collapsing magnetic field in the inductor induces a reverse electromotive force (EMF) voltage spike in the coils. The reversed polarity of the voltage spike will now allow current flow through the diode and into the corresponding lower voltage cell to charge it. This process, which is faster and more efficient than the capacitor method, is repeated until the cells are balanced. Similar to charge shuttles, this design is often only able to transfer charge between adjacent cells.

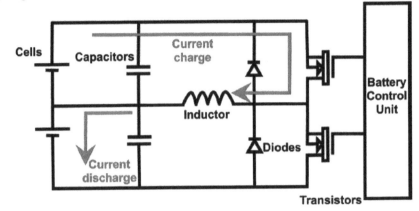

Figure 5.19 Inductive Converter

A buck boost converter is a device that can change the voltage of a direct current (DC) source to a different level. It can either increase (boost) or decrease (buck) the voltage, depending on the needs of the circuit or the load. A buck boost converter works by switching the input voltage on and off very quickly, using a transistor or a similar device. The switching creates pulses of voltage that are then smoothed by an inductor and a capacitor. The inductor stores and releases energy in the form of a magnetic field, while the capacitor stores and releases energy in the form of an electric charge. The ratio of the time that the circuit is switched on to the time that it is switched off determines the output voltage. If the switch is on for longer than it is off, the output voltage will be higher than the input voltage. If the switch is on for shorter than it is off, the output voltage will be lower than the input voltage. Unlike many DC-to-DC converters, a buck boost converter is able to both raise and lower voltage, but it also inverts the polarity of the output voltage, meaning that it has the opposite sign of the input voltage.

Flyback Transformer

A flyback transformer is a very efficient version of the inductor system, which uses the total battery pack voltage to charge a weaker cell on demand. A primary coil is charged from the full battery pack voltage, and then individual secondary coils can be switched on or off to act like a transformer and move electrical energy to a specific cell. This design will allow any weak cell in the battery pack to be charged without discharging any other individual cells, as the energy comes from the whole pack. In this design, the cells that require charging do not necessarily need to be adjacent to each other; however, it requires more space and increases circuit complexity when compared with other balancing methods.

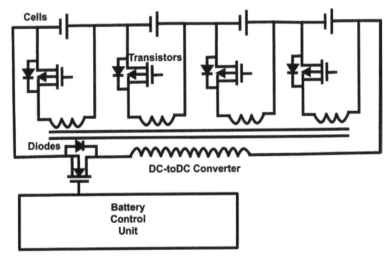

Figure 5.20 Flyback Transformer

High-Voltage Diagnosis and Testing

Lossless Balancing

A lossless active balancing method simplifies system design by reducing hardware and increasing software control. It uses a switching circuit to effectively add or remove a cell from the pack during charging and discharging by bypassing the one with the highest voltage during both processes.

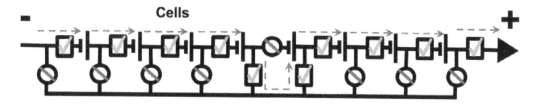

Figure 5.21 Lossless Balancing

Redox Shuttle

Chemical reactions taking place inside a battery cell with an aqueous based electrolyte prevent the cell from being overcharged.

When the cell charge reaches the maximum chemically possible, the electrolyte begins to gas, and no further charging takes place.

This process is known as redox shuttle.

If the design could be altered, redox shuttle could prevent overcharging in lithium cells by modifying the electrolyte chemistry, emulating the natural overcharge prevention seen in cells with an aqueous electrolyte.

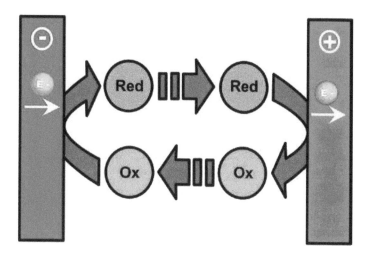

Figure 5.22 Redox Shuttle

 Redox shuttle will create internal pressures and potentially toxic and flammable gases and vapours. Although over-charge is prevented, excess current will increase gassing inside a cell, which could lead to pressures causing rupture, fire, or explosion. Charging, temperature, and pressure will need to be carefully monitored and controlled with this design to reduce potential safety issues.

High-Voltage Diagnosis and Testing

High-Voltage Battery Diagnosis and Testing

It is important to remember that batteries are chemically active and can never be shut down. However, even if the voltage potential is below the touch threshold and the risk of electrocution is minimal, other dangers still exist. Always take appropriate precautions to protect yourself from dangers.

> - Most chemicals contained within a battery, including those used to make electrodes and electrolytes, are toxic or caustic. Therefore, chemical resistant PPE is required when handling these components.
> - The process of charging or discharging may release both toxic and flammable vapours.
> - Excess temperature may cause a battery to enter a thermal runaway condition.
> - In addition, regardless of voltage being below any touch threshold, a potential difference will increase the risk of a short circuit. A short circuit may lead to arcing, sparks, and heat, which could cause injuries including electrocution, burns or impact injury from flying debris. If any combustible materials are present, this may also lead to fire or explosion. Always use certified high-voltage insulated tooling when working on or around batteries.
> - When working on or around high-voltage batteries, always wear approved/recommended HV personal protective equipment (PPE).

The diagnosis of high-voltage batteries requires in-depth testing, recording and analysis of data and results. Different test methods can be employed; however, the most beneficial information will be gathered from data and measurements when the battery is placed under load. It can be very difficult to replicate the loads placed on a high-voltage battery unless it is mounted in situ and the vehicle is being driven. A true representation of battery or cell condition can only be assessed when a circuit is switched on and current is flowing. The voltage of a switched off system, known as open circuit voltage (OCV), will always be higher than when a circuit is under load, known as closed circuit voltage (CCV). This is because voltage will only drop if current can flow.

The testing and diagnosis of high-voltage batteries will require:

- Manufacturers' data and safety information.
- Personal protective equipment (PPE), both high-voltage and chemical resistant.
- Signage and barriers to create an exclusion zone.
- Lockout protection.
- High-voltage safety rescue equipment.
- Recommended fire extinguishers.
- Insulated tooling that conforms to EN60900.
- Mechanical battery handling equipment such as lifting platforms.
- Diagnostic scan tools capable of accessing battery specifics.
- Correctly rated and calibrated electrical measurement equipment.
- Dedicated battery charging and load simulation equipment.

The following description provides an example of systematic diagnosis and testing of high-voltage batteries.

This should be viewed as one possible approach; however, safety precautions and diagnostic and testing procedures should be adapted to your particular situation.

Always refer to and follow manufacturer recommendations and procedures.

High-Voltage Diagnosis and Testing

Gather Pre-work Information and Data

Carefully question the driver about the symptoms created by the fault.

Record vehicle details such as VIN number, make and model type etc.

Gather available manufacturer data and check for any recalls if available.

Check the condition and state of charge/state of health of the low-voltage auxiliary battery.

If possible, place the vehicle into ready mode and check the driver's display for any malfunction indicator lamps (MIL).

Conduct a diagnostic scan of all available systems, read and record any diagnostic trouble codes. (Both OEM and EOBD/OBD II).
Read and record any available freeze-frame data.
Access live data relating to the high-voltage battery and electric drive system. (Test under load if possible, which may require driving the vehicle).

Compare with manufacturer specification and note any discrepancies.

Record pre-work data if available, including:

- Cell voltage values
- Internal resistance
- Temperature
- State of charge (SoC)
- State of health (SoH)
- Insulation/isolation resistance

If possible, clear diagnostic data and operate the vehicle to try and recreate the issue. (Remember that battery data will take some time to recalibrate and calculate, meaning that an extended road test may be required). Rescan for diagnostic codes and data and concentrate diagnosis around any codes that have returned.

Before you begin any diagnostic routine, you should start by carefully questioning the driver about the symptoms created by the fault. The driver will be used to their own vehicle and have first-hand experience of the fault and the symptoms it produces. Notice that I have said 'driver' and not 'customer', as these are sometimes not the same thing. The person that has brought the vehicle to the workshop may not be the main driver and any information that you get from them may be second-hand, reducing its overall usefulness.

It is important to use 'open' questions when speaking to the driver about the symptoms produced by a fault. An open question is one that does not have a simple yes or no answer.
Far more diagnostic information can be gained from asking the right type of open question and examples should include:

- Who is the main driver of the vehicle?
- What appears to be the problem/why have you brought the vehicle to the workshop?
- When did you first notice the problem?
- How often does it occur?
- What symptoms are being produced?
- Under what conditions can the problem be recreated?
- How has the vehicle been used/modified lately?

High-Voltage Diagnosis and Testing

Battery Testing

1. Set up signs and barriers to create a safety buffer zone around the vehicle.
2. Ensure that high-voltage PPE is worn as necessary.
3. Shut down the vehicle and check for absence of voltage following a systematic routine [see shutdown and isolation routine Chapter 4].
4. Gain access to and remove the battery if required. Avoid manual handling if possible and use recommended lifting equipment.
5. Prepare a battery diagnosis and test area, with warnings and restricted access.
6. Remove covers to access battery cells and modules.
7. Conduct a visual assessment of components for damage, corrosion, rupture/leakage or wear.
8. Following the manufacturer's instructions, conduct a manual assessment of cell or module voltage and record results. Remember that this will be an open circuit voltage (OCV) measurement and less reliable than a test conducted under load.
9. A cell or module with higher or lower voltage readings than others within the battery pack may provide an indication of a faulty or degraded component.
10. If a faulty cell or module is identified, it is recommended that a replacement battery pack from the original equipment manufacturer is fitted. This will ensure quality, operation and safety of the entire pack, and avoid mixing old and new cells or modules.
11. The battery pack covers or casings should be reassembled, replacing any seals or gaskets with new components. If necessary, the battery casing or shell should then be pressure tested to ensure integrity and water ingress prevention by following the manufacturer's procedures.
12. Conduct an insulation/isolation test using a megohmmeter to ensure any repairs have not caused a breach of insulation, before refitting to the vehicle.

High-Voltage Diagnosis and Testing

Notes: In order to remove a battery pack, thermal management components might need to be disconnected, which may involve draining coolant or removing refrigerant from an air conditioning system. Remember that you need to be suitably qualified to remove refrigerant from a thermal management system to conform to legal obligations.

When removing a battery, check for a centre of gravity mark, to ensure any lifting equipment is correctly aligned.
Also, be aware that once a battery has been removed, the weight distribution of vehicle components such as drive motors and so on may make the vehicle unstable on a ramp or hoist.

Safety: If you need to replace any electrical or electronic components, always ensure that the quality meets the original equipment manufacturer (OEM) specifications. If the vehicle is under warranty, using inferior parts or making deliberate modifications might invalidate the warranty. Additionally, fitting parts of inferior quality could affect vehicle performance and safety. You should only replace electrical components if the parts comply with the legal requirements for road use.

Tip: Battery cell voltage information and resistance values provided by diagnostic tool live data will be results that may include issues caused by corrosion on connectors, terminals, and components. This cumulative value may not be representative of actual battery cell/module condition. The pointed ends of a voltmeter probes can be used to penetrate corrosion, making a good contact with components, providing a more accurate result.
You should always assess the difference between values from serial data and those taken using manual measurements, as this will help you identify issues that may be caused by corrosion or connection.

Remember, however, that serial data is what a battery management system will be using for any calculations on condition and health. Therefore, this may affect the operation, safety and control of battery and drive systems.

Manual Balancing with Chargers

If a battery cell/module requires manual balancing or charging following diagnostic assessment, a specific load tester/charger designed for the type of battery should be used. Following all safety precautions, including appropriate PPE, disconnect and isolate the cell or module from the rest of the high-voltage battery pack using insulated tooling. Following the manufacturer's instructions, cycle charge/discharge to bring the voltage value of the cell or module within tolerance of the other battery cells/modules. Never leave the battery cell or module unattended while the charge/discharge procedure is being conducted. Monitor temperature and ventilate if necessary. If prismatic cells or modules are charged/discharged, never perform this action without them being correctly clamped or compressed.

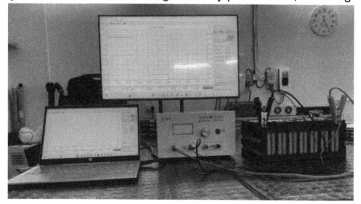

Figure 5.23 Charge/Discharge Cycling Using a Load Unit

High-Voltage Diagnosis and Testing

Post Repair Testing

1. • Following repairs and refitment, components and systems should be fully tested and results recorded.

2. • Place the vehicle in ready mode and ensure that no malfunction indicator lamps (MIL) are illuminated.

3. • Using a diagnostic tool, conduct a full scan of all available systems, to ensure no diagnostic trouble codes (DTC) are present.

4. • Access live data and compare with values checked pre and during diagnosis, repair and testing. Record results.

5. • Fully road test the vehicle and check for correct function and operation.

6. • Rescan the vehicle on all systems to ensure no permanent or pending codes have been generated during the road test.

7. • Finally, complete any necessary paperwork or documentation relating to repairs, guarantee, warranty, environmental protection and safety.

Contactors and SMR Operation

A series of relays or contactors are used to control the flow of electricity from the high-voltage battery. These relays or contactors will connect and disconnect both the positive and negative high-voltage electrical circuits in most electric vehicles. In addition to the main connection, a further relay and resistor known as a pre-charge circuit is used when initially connecting power to the system. This introduces a controlled amount of current to the system before the main relay takes over, stepping up to full voltage. Depending on manufacturer preference, the pre-charge may be found on either the positive or negative side of the high-voltage circuit.

The pre-charge relay is needed because when the high-voltage system is energised, the initial charging of the capacitors creates an inrush current which would cause excessive voltage drop at the battery cells and may lead to damage. Another issue created by the capacitor inrush current is that it can cause arcing at the contactors, which will create damage and electrical resistance, or high current might weld the contactors closed.

Resistance within the contactors will cause system operating issues because as current flows, voltage drop will occur, reducing the effectiveness of the high-voltage system. However, if the contactors weld themselves in the closed position, it will leave part of the high-voltage circuit connected. In a worst-case scenario, the high-voltage system may not be able to disconnect entirely, leaving the vehicle energised.

Figure 5.24 HV Contactors SMR

High-Voltage Diagnosis and Testing

In either situation, this will create an elevated risk of electrocution for anyone who comes into contact with the high-voltage components, especially if they assume that the high-voltage system has shut down and disconnected. To help combat this, the contactors or system main relays switch on and off with a systematic routine. The drive management system checks for high-voltage during the power up and power down sequence, and if an unexpected voltage is found, the process is halted, and a diagnostic trouble code is set. A warning will be shown on the driver display panel and may prevent the vehicle from entering ready mode and being operated.

Regarding terminology, there is a difference between relays and contactors, but the terms are often used interchangeably as they are essentially performing the same task of switching a circuit.

A relay will often handle relatively low currents and have a single switched contact that can be operated at any one time. (Relays may have multiple switched contacts, but normally only one is operated at a time). Relays can be designed as either normally open or normally closed.

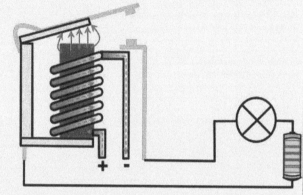

A contactor, on the other hand, is designed to connect two terminals in a circuit with normally open contacts. Because contactors typically carry high loads, they often have additional safety features, such as spring-loaded contacts, to help ensure the circuit is broken when de-energised. This is important because, in high-load situations, current flow and arcing may cause the contacts to weld themselves together, resulting in a situation where the system cannot be de-energised. Contactors may also have arc suppression features.

Magnetic arc suppression works by extending the path an arc would travel, spreading it out towards the arc chamber walls. If the distance is extended further than the electrical energy can overcome, the arc is suppressed. This can be enhanced by adding a gas to the arc chamber, which helps to rapidly cool any arcing that occurs. If the arc chamber is encapsulated in ceramic, the gas may be hydrogen. An absence of oxygen ensures that the hydrogen gas helps spread the arc quickly towards the chamber walls where it rapidly cools arcing but does not form a flammable compound.

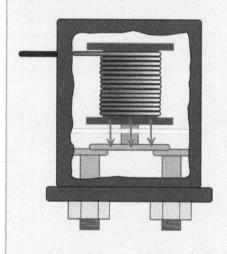
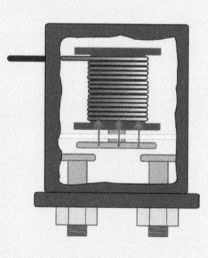

High-Voltage Diagnosis and Testing

Capacitor Operation

Capacitors are used in the high-voltage system of an electric vehicle as a temporary storage of electrical energy that can be quickly accessed by the drive motors.
They act as a 'buffer' or 'shock absorber' between the battery and motor generators.
The large energy demand from the drive motors may cause an excessive volt drop at the high-voltage battery, with potential to create irreversible cell damage. Instead, the large energy demand can be sourced from the capacitors, which have no chemical compounds that are affected by the volt drop; the battery only has to supply enough electricity to keep the capacitors topped up.

When generators operate during a charge sequence, alternating current (AC) is first rectified to direct current (DC) and then passed through the capacitors before returning to the battery. The capacitors will both smooth any residual AC ripple and help provide a minimal form of energy regulation between the generators and the battery.

When the vehicle is switched off, the capacitors could store a potentially lethal amount of electrical energy for a considerable amount of time. As a result, every time the vehicle is powered down, the capacitors must be emptied in order to deenergise the high-voltage system for safety.

Two methods are used to discharge capacitors: active and passive.

Active discharge occurs when the vehicle is switched off by allowing the stored energy in the capacitors to flow into the stator coils of the electric drive motors. Stored capacitor electricity is direct current (DC) and the drive motors operate using alternating current (AC), so no torque is created, but energy is dissipated as heat in approximately one second.

Passive discharge is needed to provide a safety backup method to empty the capacitors in case the active system fails. A high-ohmic resistor is placed in parallel across the positive and negative connections of the capacitor circuit. When the system is switched on and the contactors are closed, electricity takes the path of least resistance, disregarding the high-ohmic resistor and filling the capacitors. When the vehicle is switched off, and the contactors open, there is now a circuit created by the resistor between the positive and negative terminals of the capacitor. The capacitors discharge their stored energy through the resistor, which is then converted into heat. This process will often take approximately five to ten minutes to bring the stored voltage below the touch threshold.

A capacitor is a component that acts as a temporary storage of electrical energy. Unlike a battery cell, it does not create its own electricity through chemical reaction, but instead must be charged.
A capacitor could be likened to a bucket for electricity, which can be filled or emptied on demand. They are used as a reservoir to smooth out the peaks and troughs of electrical system demands.

Active capacitor discharge through a drive motor is like emptying the bucket down the drain.
Passive capacitor discharge is like drilling a small hole in the bucket, allowing the contents to leak away over time.

When a vehicle is initially switched on and powered up, the capacitors (buckets) are empty and must be filled. This is similar to priming a fuel system on an internal combustion engine vehicle by running the fuel pump for a few seconds. The empty capacitors are very greedy and current flow from the battery could cause cell damage or weld system main relays (SMR) closed. As a result, a process to control the charging of capacitors is needed, similar to a 'cold start'. This is the purpose of the pre-charge circuit. This process is like filling the buckets gradually by turning the tap on halfway.

High-Voltage Diagnosis and Testing

Pre-charge Circuits

A pre-charge system is connected to either the positive or negative main battery connection circuit. It uses a relay and resistor that bypasses one of the system main relays. When the start button is pressed, the pre-charge relay is energised, along with its corresponding SMR. The pre-charge relay directs the initial flow to the capacitors through a resistor, limiting the current flow. Once the initial capacitor demand has reached an acceptable level, the other SMR is switched on and the pre-charge is switched off. This whole procedure happens in a few hundred milliseconds.

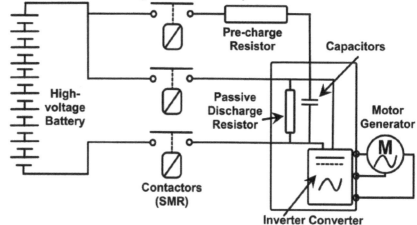

Figure 5.25 SMR Pre-charge and Capacitor Circuit

Start Up and Shut Down

To ensure a safety check is conducted on the contactors or system main relays, a power up and power down sequence will engage/disengage them in a specific order to detect that none are welded closed. It does this by monitoring for unexpected voltage at the capacitors and in the inverter unit. The sequence will vary depending on whether the pre-charge is on either the positive or negative circuit.

Positive pre-charge circuit sequence:

- When the start button or ignition is switched on, the diagnostic system first checks for any stored codes relating to the SMR or contactor system. If a code is detected, start-up is halted, and a malfunction indicator light (MIL) is illuminated. If no preexisting codes are present, the start-up procedure continues.
- The pre-charge relay is energised, and a voltage check is conducted. If a voltage increase is detected at the capacitors or inverter, this indicates that the negative SMR is welded. The start-up is halted, a code is set, and the malfunction indicator light (MIL) is illuminated. If no voltage increase is detected, the pre-charge relay is switched off and the start-up process continues.
- The main negative SMR or contactor is now switched on and a voltage check is conducted. If a voltage increase is detected at the capacitors or the inverter, this indicates that the pre-charge relay is welded. The start-up is halted, a code is set, and the malfunction indicator light (MIL) is illuminated. If no voltage increase is detected, the start-up process continues.
- The pre-charge is switched back on, and the initial charging of the capacitors takes place.
- Once the initial inrush current to the capacitors has happened, the main positive SMR or contactor is energised, bringing the system to full operating voltage, and the pre-charge relay/circuit is disengaged.
- A check on main system operating voltage is now conducted to ensure that it is within the manufacturer's specification. If the main circuit voltage is too low, start-up is halted to protect the drive and propulsion system, a code is set, and the malfunction indicator light (MIL) is illuminated. If main circuit voltage is within tolerance, the vehicle will enter ready mode.
- When the vehicle is switched off, the shutdown process begins.
- All load is firstly removed from the high-voltage system to prevent current flow through the system main relays or contactors.
- The positive SMR or contactor is switched off first in conjunction with active capacitor discharge and a voltage check is conducted. If voltage does not fall/reduce, this indicates that the positive SMR is welded. A code is set, and the malfunction indicator light (MIL) is illuminated. As long as the voltage falls, the shutdown continues.
- When voltage has fallen below a preset value, the negative SMR is switched off, and shutdown has finished.

High-Voltage Diagnosis and Testing

>
>
> A summary of the start-up/shutdown sequence for a positive pre-charge is shown below:
> 1. Power/ignition on
> 2. Pre-charge on
> 3. Pre-charge off
> 4. Negative SMR on
> 5. Pre-charge back on
> 6. Positive SMR on
> 7. Pre-charge off
> 8. Ready mode engaged
> 9. Power/ignition off
> 10. Positive SMR off
> 11. Negative SMR off
> 12. Finish
>
> - ☑ Negative SMR weld codes are created during power up.
> - ☑ Pre-charge weld codes are created during power up.
> - ☑ Capacitor charge time too long codes are created during power up.
> - ☑ Main system voltage too low codes are created in ready mode.
> - ☑ Positive SMR weld codes are created during power down.
>
> (Two weld checks are conducted during power up, and one check is conducted during power down.)

Negative pre-charge circuit sequence:

- When the start button or ignition is switched on, the diagnostic system first checks for any stored codes relating to the SMR or contactor system. If a code is detected, start-up is halted, and a malfunction indicator light (MIL) is illuminated. If no preexisting codes are present, the start-up procedure continues.
- The main positive SMR or contactor is switched on first and a voltage check is conducted. If a voltage increase is detected at the capacitors or the inverter, this indicates that the pre-charge relay is welded. The start-up is halted, a code is set, and the malfunction indicator light (MIL) is illuminated. If no voltage increase is detected, the start-up process continues.
- The pre-charge is now switched on, and the initial charging of the capacitors takes place.
- Once the initial inrush current to the capacitors has happened, the main negative SMR or contactor is energised, bringing the system to full operating voltage, and the pre-charge relay/circuit is disengaged.
- A check on main system operating voltage is now conducted to ensure that it is within the manufacturer's specification. If the main circuit voltage is too low, start-up is halted to protect the drive and propulsion system, a code is set, and the malfunction indicator light (MIL) is illuminated. If main circuit voltage is within tolerance, the vehicle will enter ready mode.
- When the vehicle is switched off, the shutdown process begins.
- All load is firstly removed from the high-voltage system to prevent current flow through the system main relays or contactors.
- The negative SMR or contactor is switched off first in conjunction with active capacitor discharge and a voltage check is conducted. If voltage does not fall/reduce, this indicates that the negative SMR is welded. A code is set, and the malfunction indicator light (MIL) is illuminated. As long as the voltage falls, the shutdown continues.
- When voltage has fallen below a preset value, the positive SMR is switched off, but to check that the positive SMR has not welded, the pre-charge is switched back on.
- The pre-charge relay is energised, and a voltage check is conducted. If a voltage increase is detected at the capacitors or inverter, this indicates that the positive SMR is welded. A code is set, and the malfunction indicator light (MIL) is illuminated. If no voltage increase is detected, the pre-charge relay is switched off and shutdown has finished.

High-Voltage Diagnosis and Testing

A summary of the start-up/shutdown sequence for a negative pre-charge is shown below:
1. Power/ignition on
2. Positive SMR on
3. Pre-charge on
4. Negative SMR on
5. Pre-charge off
6. Ready mode engaged
7. Power/ignition off
8. Negative SMR off
9. Positive SMR off
10. Pre-charge on
11. Pre-charge off
12. Finish

- ☑ Pre-charge weld codes are created during power up.
- ☑ Capacitor charge time too long codes are created during power up.
- ☑ Main system voltage too low codes are created in ready mode.
- ☑ Negative SMR weld codes are created during power down.
- ☑ Positive SMR weld codes are created during power down.

(One weld check is conducted during power up, and two checks are conducted during power down.)

Contactor and SMR Diagnosis and Testing

If a vehicle fails to start, or enter ready mode, several steps should be taken before suspecting a fault with the contactors and system main relays.

1. Check to see if the low-voltage auxiliary battery is at an acceptable state of charge by measuring its voltage.

2. As many vehicles require the operator to place their foot on the brake in order to engage ready mode, check to make sure that the brake lights and brake light circuit are working correctly.

3. Connect a diagnostic scan tool to the data link connector and conduct a full system scan for any pending or present diagnostic codes.

If any diagnostic codes relate to a potential failure with the contactor or SMR circuits, the vehicle should be shut down and isolated [following the procedures described in Chapter 4]. The vehicle can then be dismantled to gain access to the contactors and system main relays to enable testing to take place.

Testing contactors or system main relays will often require the battery or high-voltage system covers to be removed. These covers are designed to prevent accidental contact with exposed high-voltage connections. Sometimes, high-voltage interlock loops are used as additional protection, and these may require bypassing in order for any testing to take place. Be aware that bypassing any safety system means that exposed live electrical connections can easily cause injury or death.

High-Voltage Diagnosis and Testing

Diagnostic testing of 'live' systems should only be conducted if you have received adequate training, using the correct personal protective equipment (PPE), fully insulated tooling, and correctly rated and calibrated diagnostic electrical test equipment.

The following diagnostic descriptions are designed to support knowledge and understanding, but do not act as a substitute for appropriate training. Never attempt any diagnosis or repairs unless you are suitably qualified and have the correct tools, equipment, and safety measures in place. When conducting diagnosis on 'live' high-voltage systems, try to avoid lone working.

Resistance Testing Contactors and SMR

Once the high-voltage battery has been isolated and capacitors discharged, conduct a test to confirm the absence of voltage.
To correctly test resistance using an ohmmeter, the contactors or system main relays must be fully isolated, powered down, and disconnected from the rest of the circuit.

1. Set the ohmmeter to the appropriate scale and calibrate it by touching the two probes together. Subtract any resistance shown on the display from any final results.

2. Connect the probes of the ohmmeter across the high-voltage connections of each contactor or system main relay in turn.

3. With the contactors or relays not energised, a reading of off/out of limits (O/L) or infinity should be obtained. (A low resistance result may indicate that the contactors or system main relays are welded closed).

4. Record the results and compare them with the manufacturer specification.

5. Energise the contactors or system main relays by connecting them to an appropriate low-voltage power source.

6. Repeat the resistance test across the high-voltage connections of each contactor or system main relay and record the results.

7. Compare these results with the manufacturer specification. (The lower the resistance reading, the better).

When testing the resistance of the relay and resistor on the pre-charge circuit, the specification of the resistor is often indicated on the resistor itself.

Remember that resistance tests using an ohmmeter are less reliable because the test is conducted with no current flowing in the circuit. Also, the low voltage battery contained inside the ohmmeter itself, which is used as part of the resistance calculation, is not representative of the voltages used in the drive system of an electric vehicle.

High-Voltage Diagnosis and Testing

 You will need access to manufacturer specific data relating to expected resistances to assess the serviceability of contactors or system main relays when using this form of test.

Volt Drop Testing Contactors and SMR

 A volt drop test is far superior at assessing resistance within the high-voltage contactor or system main relays. However, to conduct a volt drop test, the vehicle will need to be powered up and placed in ready mode, meaning that the high-voltage connections are exposed and live. Do not conduct this test without appropriate training, insulated hand tools, correctly rated and calibrated equipment, and high-voltage personal protective equipment (PPE).

To conduct a volt drop test on the contactors or system main relays, you will need a correctly rated and calibrated voltmeter. This should be a minimum of category 3 (CAT III) 1000 V. Where possible, use crocodile clips or back probe connections so that you do not have to handle the probes of the voltmeter while the system is powered up.

1. With the contactors and system main relays in situ, connect the voltmeter across the high-voltage terminals of each contactor individually. For example, positive circuit and then negative circuit.

2. Power up the vehicle and place it in ready mode to energise the contactors.

3. Voltage results shown should be as close as possible to zero. Any voltage shown on the voltmeter display indicates a high resistance across the switched terminals inside the system main relays or contactors.

 The simple act of connecting a voltmeter across the terminals of a contactor or system main relay provides an alternative pathway for electric current to flow and acts like a jumper lead. This means as soon as the voltmeter is connected across the contactors or system main relay, current can flow and begin to charge the capacitors at the inverter. Even though the system may not yet be powered up or placed in ready mode, voltage potential and current are now present, increasing the risk of electrocution, injury, or death. Therefore, you must take appropriate precautions, such as wearing high-voltage personal protective equipment (PPE), using insulated hand tools, and following the manufacturer's guidelines.

Try to make voltmeter probe connections/disconnections one at a time as this will help reduce the possibility of making two points of contact with your hands to a high-voltage circuit.

When placing the vehicle in ready mode, try to remain fully inside the vehicle so that you are not creating a potential difference between two parts of your body.

High-Voltage Diagnosis and Testing

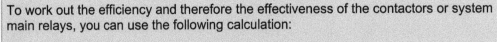

The results of a voltage drop test across the terminals of the system main relays or contactors can help provide an indication of their efficiency in proportion to the resistance.

To work out the efficiency and therefore the effectiveness of the contactors or system main relays, you can use the following calculation:

Voltage drop/System voltage x 100 = Efficiency loss of the contactors or system main relays

Example:
Voltage drop = 26 volts
System voltage = 360 volts

26/360 x 100 = 7.22% This is the efficiency loss of the contactors or system main relays in this example.

Oscilloscope Testing Contactors and SMR

Positive Pre-charge Circuit

To test the contactors and system main relays of a positive pre-charge system circuit, you should use three channels. You will need to access the low-voltage feed wires of the contactors or SMRs and either back probe them or connect them using breakouts.

KEY:

- Channel 1 shows the main positive contactor or SMR
- Channel 2 shows the positive pre-charge contactor or SMR
- Channel 3 shows the main negative contactor or SMR

The following image shows the oscilloscope readings for each channel during the start-up, ready mode, and shutdown sequences of a positive pre-charge system circuit. The image also shows the points on the waveform where the contactors or SMRs are switched on or off, and the checks for possible welding.

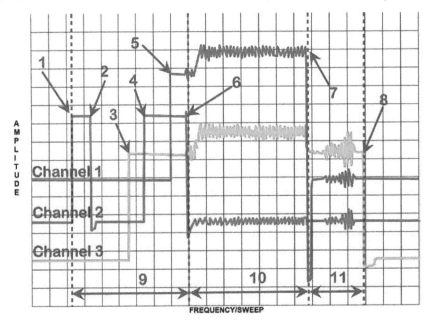

Figure 5.26 Positive Pre-charge Circuit Waveform

High-Voltage Diagnosis and Testing

Table 5.5 Positive pre-charge circuit contactors and SMR	
Waveform component	**Description**
1	**Channel 2.** This is the point on the waveform where the positive pre-charge SMR is switched on. This checks to see if the main negative contactor or SMR is welded.
2	**Channel 2.** This is the point on the waveform where the positive pre-charge SMR is switched off.
3	**Channel 3.** This is the point on the waveform where the main negative contactor or SMR is switched on. This checks to see if the pre-charge SMR is welded.
4	**Channel 2.** This is the point on the waveform where the positive pre-charge SMR is switched back on to charge the capacitors.
5	**Channel 1.** This is the point on the waveform where the main positive contactor or SMR is switched on.
6	**Channel 2.** This is the point on the waveform where the positive pre-charge SMR is switched off and the start-up sequence has finished.
7	**Channel 1.** This is the point on the waveform where the main positive contactor or SMR is switched off. This checks to see if the main positive contactor or SMR is welded.
8	**Channel 3.** This is the point on the waveform where the main negative contactor or SMR is switched off and the shutdown sequence has finished.
9	This section indicates the start-up sequence and period. The start-up sequence begins when the vehicle is powered up and ends when the contactors and SMRs are fully energised, and the capacitors are charged. The start-up sequence should take less than a second.
10	This section indicates the period of ready mode. The ready mode begins when the start-up sequence has finished and ends when the vehicle is powered down. The ready mode should last as long as the vehicle is in operation.
11	This section indicates the shutdown sequence and period. The shutdown sequence begins when the vehicle is powered down and ends when the contactors and SMRs are fully de-energised, and the capacitors are discharged. The shutdown sequence should take less than a second.

Negative Pre-charge Circuit

To test the contactors and system main relays of a negative pre-charge system circuit, you should use three channels. You will need to access the low-voltage feed wires of the contactors or SMRs and either back probe them or connect them using breakouts.

KEY:

- Channel 1 shows the main negative contactor or SMR
- Channel 2 shows the negative pre-charge contactor or SMR
- Channel 3 shows the main positive contactor or SMR

High-Voltage Diagnosis and Testing

The following image shows the oscilloscope readings for each channel during the start-up, ready mode, and shutdown sequences of a negative pre-charge system circuit. The image also shows the points on the waveform where the contactors or SMRs are switched on or off, and the checks for possible welding.

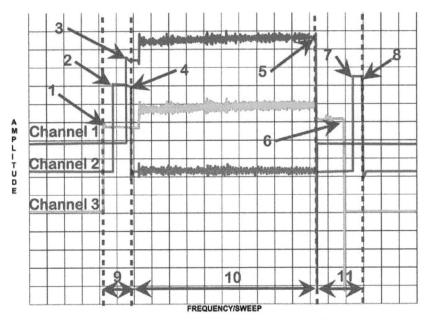

Figure 5.27 Negative Pre-charge Circuit Waveform

Table 5.6 Negative pre-charge circuit contactors and SMR	
Waveform component	**Description**
1	**Channel 3.** This is the point on the waveform where the main positive contactor or SMR is switched on. This checks to see if the pre-charge SMR is welded.
2	**Channel 2.** This is the point on the waveform where the negative pre-charge SMR is switched on to charge the capacitors.
3	**Channel 1.** This is the point on the waveform where the main negative contactor or SMR is switched on.
4	**Channel 2.** This is the point on the waveform where the negative pre-charge SMR is switched off and the start-up sequence has finished.
5	**Channel 1.** This is the point on the waveform where the main negative contactor or SMR is switched off. This checks to see if the main negative contactor or SMR is welded.
6	**Channel 3.** This is the point on the waveform where the main positive contactor or SMR is switched off.
7	**Channel 2.** This is the point on the waveform where the negative pre-charge SMR is switched back on to check if the main positive contactor or SMR is welded.
8	**Channel 2.** This is the point on the waveform where the negative pre-charge SMR is switched off and the shutdown sequence has finished.

High-Voltage Diagnosis and Testing

Table 5.6 Negative pre-charge circuit contactors and SMR	
9	This section indicates the start-up sequence and period. The start-up sequence begins when the vehicle is powered up and ends when the contactors and SMRs are fully energised, and the capacitors are charged. The start-up sequence should take less than a second.
10	This section indicates the period of ready mode. The ready mode begins when the start-up sequence has finished and ends when the vehicle is powered down. The ready mode should last as long as the vehicle is in operation.
11	This section indicates the shutdown sequence and period. The shutdown sequence begins when the vehicle is powered down and ends when the contactors and SMRs are fully de-energised, and the capacitors are discharged. The shutdown sequence should take less than a second.

To use an oscilloscope to substantiate the results of diagnostic trouble codes (DTCs) relating to system main relays and contactors, you will need to connect the oscilloscope to the low-voltage circuit.

If a fault exists in the system main relay start-up circuits, the vehicle will not power up or enter ready mode and a malfunction indicator light (MIL) should be illuminated on the driver's display. To allow the start-up sequence to run, you will need to clear any DTCs from the system memory. When you press the start button, the start-up sequence will begin. If a fault is detected, the waveforms on the oscilloscope will flatline at the point of failure, allowing you to confirm the DTC.

If a fault exists in the system main relay shutdown circuits, the vehicle will power down, but should finish with a MIL illuminated on the driver's display the next time the vehicle is powered up. To allow the start-up and shutdown sequence to run, you will need to clear any DTCs from the system memory. When you press the start button to switch the vehicle off, if a fault is detected, the waveforms on the oscilloscope will flatline at the point of failure, allowing you to confirm the DTC.

You will only get one opportunity between clearing codes to record and analyse waveforms.

Vehicle Operating Information and CAN Bus

Vehicle operating information is shared between different systems using an in-vehicle network. In a network, the control units, often referred to as nodes, are linked by an in-vehicle communication system that allows the transmitting and receiving of data. CAN Bus is probably one of the most widely used networks within vehicle design, and the name 'CAN Bus' has become synonymous with ECU communication to the point where it is often used to describe all in-vehicle networking, even if another type is actually being used.
Controller area network (CAN) was introduced by Robert Bosch in the 1980s and is an international standards organisation (ISO) standard for a serial multiplex communication protocol. CAN Bus is a network communication standard where information is bundled into a 'data packet' and sent onto the bus system along two twisted wires. Every node on the bus system receives the message and acts if required.
The advantages of CAN Bus are:

> Transmission speeds are much faster than those used in conventional communication (up to 1 Mbps), allowing much more data to be sent.
> The system is very immune to interference (noise), and the data obtained from each error detection device is more reliable.
> Each ECU connected via the CAN Bus communicates independently, therefore if an ECU is damaged or faulty, communications can be continued in many cases.

High-Voltage Diagnosis and Testing

Physical Layer of In-Vehicle Network Systems

The physical layer is the name given to the wiring of an in-vehicle network system. It is used to join the various nodes or components to each other, and also to connect networks of different speeds and systems. To help locate and trace the network systems, manufacturers create wiring schematics known as topology diagrams, showing the layout of the major components.

The CAN Bus line consists of two cables, known as CAN H and CAN L (CAN High and CAN Low, respectively). The CAN High and Low wires are twisted together, and this helps to cancel out noise which may be caused by electromagnetic interference from other vehicle electrical systems. At the ends of each CAN line are terminal resistors that help to dampen out voltage spikes (back EMF) which could be caused as the communication is triggered on and off. The CAN Bus lines connecting two dominant ECUs are the main bus lines, and the CAN Bus lines connecting each individual ECU are the sub-bus lines. Each ECU communicates with the CAN Bus periodically, sending information from several sensors. This information is circulated on the CAN Bus as a data packet. Each ECU needing data on the CAN Bus can receive these data frames sent from each ECU simultaneously. A single ECU transmits multiple data frames. When data packets conflict with one another (when more than one ECU transmits signals at the same time), data is prioritised for transmission by a process called 'mediation'.

If mediation is required:

1. The data frame with high priority is transmitted first according to ID codes embedded in the data packet.
2. Transmission of low-priority data is suspended by the issuing ECU until the bus clears (when no transmission data exists on the CAN Bus).
3. The ECU containing suspended data frames transmits when the bus becomes available.

Communication Data of In-Vehicle Network System

When an ECU receives a signal from a vehicle sensor, it processes it and places the information on the network bus as a data packet. The data packet is usually made up of the following components:

- A header, **SOF (Start of Frame)**: the equivalent of 'hello, I am transmitting a message'.
- The priority **ID (Identifier)** region: how important this message is.
- Remote transmission request **RTR**: indicates whether a message is a data frame or a remote request frame. A data frame contains actual data, while a remote request frame asks for data from another ECU.
- Flexible data frame **FDF**: indicates whether a message is a CAN FD frame or a Classical CAN frame. CAN FD is an extension of the CAN protocol that allows for higher data rates and longer data fields.
- Data length **Control region DLC**: this tells the receiver how many bytes of data are in the packet.
- Data type **Data region**: this indicates what type of information is contained, e.g. voltage, speed, temperature, etc.
- Data **Data region**: the actual sensor information itself.
- An error detection code **Cyclic Redundancy Check (CRC)** region: this verifies that all the information has been received correctly.
- End of message **EOF (End of Frame)**: 'goodbye'.
- Finally, a request for a response from the receiving ECU **ACK (Acknowledge) region**: this says, 'thank you, I got your message'.

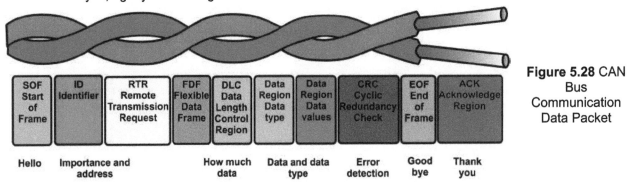

Figure 5.28 CAN Bus Communication Data Packet

High-Voltage Diagnosis and Testing

Reducing Data Corruption in CAN Bus Systems

To help reduce the possibility of data corruption caused by misinterpretation or external electromagnetic interference, a CAN Bus system uses two communication wires instead of one, twisted over and over each other in a spiral.

The same data is sent on both of these communication wires as an on and off voltage signal.

One signal is sent as a positive switch and one is sent as a negative switch, providing a mirror image on each network wire, which are known as CAN High and CAN Low.

The potential difference between the voltages on the two lines produces a digital signal that can be processed into information.

Data Transmission - High Speed

The transmitting ECU sends switched voltage through the CAN H and CAN L bus.

It sends 2.5 to 3.5 volt signals to the CAN High line and 2.5 to 1.5 volt signals to the CAN Low line.

The receiving ECU reads the data from the CAN lines as a potential difference of between 3.5 and 1.5 volts.

In **Figure 5.29**, 'Recessive' refers to the state where both CAN H and CAN L are at the 2.5 volt state, and 'Dominant' refers to the state where CAN H is at the 3.5 volt state and CAN L is at the 1.5 volt state. These values correspond to a binary value of either 1 or 0.

Recessive = Logic value of 1
Dominant = Logic value of 0

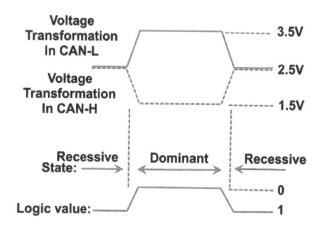

Figure 5.29 Data Transmission High Speed

Data Transmission - Low Speed

The transmitting ECU sends switched voltage through the CAN H and CAN L bus.

It sends 0 to 4 volt signals to the CAN High line and 1 to 5 volt signals to the CAN Low line.

The receiving ECU reads the data from the CAN High and CAN Low as a potential difference of between 5 and 0 volts.

In **Figure 5.30**, 'Recessive' refers to the state where CAN H is at 0 volt and CAN L is at 5 volt, and 'Dominant' refers to the state where CAN H is at 4 volt and CAN L is at 1 volt.

Recessive = Logic value of 1
Dominant = Logic value of 0

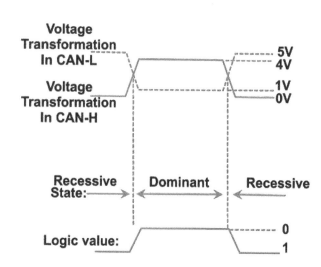

Figure 5.30 Data Transmission Low Speed

High-Voltage Diagnosis and Testing

Termination resistances can give a good indication of correct circuit operation. If an ohmmeter is connected in parallel across CAN High and CAN Low (using pins 6 and 14 of the data link connector for example), then the total recorded resistance will be halved.

- If 60Ω is shown, CAN High and CAN Low should be OK.
- If O/L (infinity) is shown, an open circuit exists in both lines.
- If 0Ω is shown, a dead short exists.
- If 120Ω is shown, one CAN line may be at fault (confirm communication using an oscilloscope).

Network Diagnosis of In-Vehicle Network Systems

If a critical network failure occurs, such as a short to positive or ground, the vehicle may suffer a complete communication loss. With a networked system, if communication is lost within a certain area, a number of items will not work, and numerous trouble codes may be generated.

Having connected a scan tool and retrieved the diagnostic trouble codes, you should look for the code that is the root cause. Communication failures are normally an effect of the original fault (i.e. 'unable to communicate' or 'communication lost'). You should ask yourself, 'Is this the cause or an effect created by the fault?' CAN Bus systems report communication faults as live data. As a result, once you have identified the causal trouble code, you may be able to conduct a diagnosis by disconnecting and isolating components or sections of the low-voltage wiring loom until communication is re-established.

It will often be necessary to connect an oscilloscope to help further investigate a possible network fault. The following section shows CAN Bus network communication and how to interpret the waveforms produced.

Oscilloscope Testing of CAN Bus

A CAN system can often be identified as a pair of twisted wires entering or leaving an ECU. An oscilloscope can be connected to these wires by 'back probing' at the ECU socket, but if the vehicle is operating using CAN Bus, a good place to connect to the main circuit is at pins 6 and 14 of the diagnostic socket (DLC).

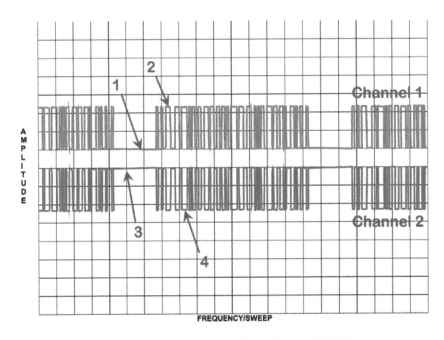

Figure 5.31 CAN Bus (CAN H and CAN L)

High-Voltage Diagnosis and Testing

Table 5.7 Waveform analysis CAN High and CAN Low	
Waveform component	**Description**
1	**Channel 1** is connected to CAN H (High) and switches positively. This means that the voltage is 0 or 2.5 volts in the off position depending on network speed.
2	**Channel 1**. When switched on, the voltage will jump to 3.5 or 4 volts depending on network speed.
3	**Channel 2** is connected to CAN L (Low) and switches negatively. This results in a voltage of 5 or 2.5 volts in the off position depending on network speed.
4	**Channel 2**. When switched on, the voltage will fall to 1 or 1.5 volts depending on network speed.

By changing the frequency/sweep on the oscilloscope and aligning the voltage amplitudes between Channel 1 and Channel 2, it is possible to compare the two patterns and see the potential difference from CAN High and CAN Low. This can then be interpreted as a dominant or recessive logic value. [See **Figures 5.29** and **5.30**].
Recessive = Logic value of 1
Dominant = Logic value of 0

It is important to check that the patterns from CAN High and Low show equal and opposite with clean edges when examining the waveform. This indicates that the network wiring circuit is operating effectively, and that any non-responsive individual ECU is likely caused by the ECU itself.

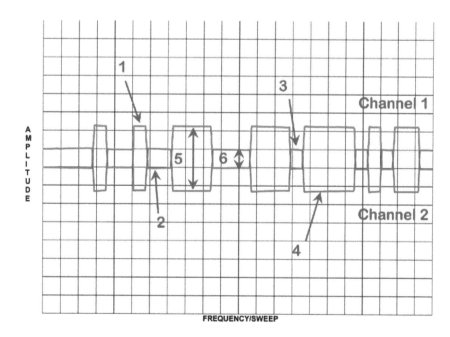

Figure 5.32 CAN Bus Potential Difference

High-Voltage Diagnosis and Testing

Waveform component	Table 5.8 Waveform analysis CAN Bus potential difference
	Description
1	**Channel 1** is connected to CAN H (High) and switches positively. This means that the voltage is 3.5 or 4 volts in the on position depending on network speed.
2	**Channel 1.** When switched off, the voltage will fall to 0 or 2.5 volts depending on network speed.
3	**Channel 2** is connected to CAN L (Low) and switches negatively. This results in a voltage of 5 or 2.5 volts in the off position depending on network speed.
4	**Channel 2.** When switched on, the voltage will fall to 1 or 1.5 volts depending on network speed.
5	This section of the waveform shows a dominant logic value of 0.
6	This section of the waveform shows a recessive logic value of 1.

It is often possible to conduct an initial oscilloscope diagnosis of network systems at the pins of the vehicle data link connector. Due to the standardised layout of the 16-pin connector the terminals can be identified from the image shown below:

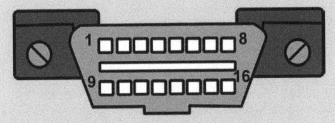

1. Manufacturer specific [sometimes used for network communication].
2. Bus positive SAE J1850 PWM and VPW.
3. Manufacturer specific [sometimes used for network communication].
4. Chassis ground.
5. Signal ground.
6. CAN High.
7. K-Line of ISO9141-2 and ISO14230-4.
8. Manufacturer specific [sometimes used for network communication].
9. Manufacturer specific [sometimes used for network communication].
10. Bus negative SAE J1850 PWM.
11. Manufacturer specific [sometimes used for network communication].
12. Manufacturer specific [sometimes used for network communication].
13. Manufacturer specific [sometimes used for network communication].
14. CAN Low.
15. L-Line of ISO9141-2 and ISO14230-4.
16. Battery voltage.

Using insulation piercing probes to measure CAN Bus signals is not recommended, because this can damage the integrity of the wiring and promote communication problems.

High-Voltage Diagnosis and Testing

Motor Generator Operation

The drive motors of an electric vehicle are often synchronous alternating current (AC) types. They consist of a three-phase wire-wound stator and a permanent magnet rotor, [for the construction and operation of different motor types, see Chapter 4]. An alternating current is fed to each of the three phases in sequence, providing alternating positive and negative magnetic fields, which can be used to attract or repel the rotor so that it turns. The strength of the rotating magnetic field dictates the torque provided, and the frequency dictates its speed. Motor generators are normally encased within the transmission housing, and are difficult to access; however, diagnostic testing is possible, often with the motor generator in situ.

Motor Generator Testing

Phase Resistance

The resistance of the wire wound phases in the stator of a motor generator can provide an indication of motor generator condition. A specialist resistance measuring tool known as a milliohm meter will be required for this operation. This is a form of highly sensitive electrical test equipment, designed to measure very small amounts of electrical resistance. Its connections to the electrical circuit will often use 'Kelvin clips', which are a type of electrical clip that provides superior grip with insulated jaws. It will also often have some form of temperature measurement and compensation device to overcome potential miscalculations and results created by heat.

Figure 5.33 A Milliohm Meter

Once the high-voltage battery has been isolated and capacitors discharged, conduct a test to confirm the absence of voltage.
To correctly test phase resistance using a milliohm meter, the motor generator must be fully isolated, powered down and disconnected from the rest of the circuit.

This can often be achieved by disconnecting the three phase wires at the inverter unit. However, the motor generator can often be left in situ.

1. Any temperature compensation probe must be mounted as close as possible to the stator of the motor generator.

2. Set the milliohm meter to the appropriate scale and calibrate it by touching the probes or clips together.

3. The phases of a motor generator unit are often labelled U, V, and W. It is important to compare the resistance between all three of the phases.

4. Systematically connect the clips of the milliohm meter and record the results. For example, connect phases U to V, V to W and W to U.

5. Allowing for temperature compensation, the results from each of the three phases should be very similar. Any significant variation in resistance may indicate a faulty motor generator. You may require access to manufacturer-specific data in order to accurately assess any resistance tolerance.

High-Voltage Diagnosis and Testing

Milliohm meters and motor generator testers are available that are specifically designed to conduct automatic testing when connected to the motor generator circuit's phases.

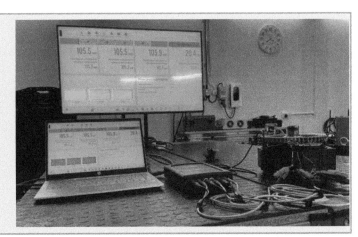

A milliohm meter can also be used to check the condition of equipotential bonding (EPB). With the wiring loom disconnected, the milliohm meter can be attached to either end of the bonding, and a measurement taken. A maximum of 100 milliohms per meter of equipotential bonding is often satisfactory, but please refer to manufacturer's specifications.

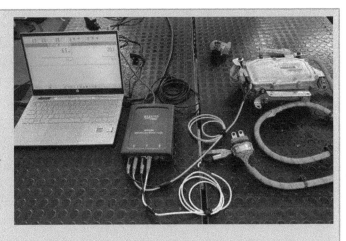

Oscilloscope Voltage and Current

Using an oscilloscope for testing the voltage and current of motor generators may require the exposure of live circuits. This means the risk or danger of electrocution is greatly increased. Also, testing motor generators using an oscilloscope may not be suitable unless the equipment is specifically rated for use with a high-voltage system. Specially designed breakout leads may be required in order to access voltage and current connections.

Remember that even though MILs and codes may not be present, faults may exist which have not been detected by the vehicles diagnostic system.

Connecting breakout leads will often disconnect equipotential bonding (EPB), bypassing this safety system and increasing the risk of electrocution should something go wrong.

Following any diagnostic test, all systems should be fully/correctly reassembled and tested for operation and safety.

High-Voltage Diagnosis and Testing

Current Test – Oscilloscope

This test is performed using a non-intrusive current clamp and a set of specifically designed three-phase breakout leads.

1. Once the high-voltage battery has been isolated and capacitors discharged, conduct a test to confirm the absence of voltage.

2. Connect the breakout leads and reassemble the high-voltage system, including any covers or safety devices used for operator protection.

3. Wearing appropriate high-voltage PPE, connect an inductive clamp (or multiple clamps if available) to the breakout phase wire(s).

4. Power the vehicle into ready mode, check malfunction indicator lights (MIL), and conduct a diagnostic scan to ensure no safety issues have been detected by the vehicle.

5. With the vehicle raised and correctly supported with the driven wheels off the ground, select a drive mode or rotate the wheels by hand and note the waveforms produced on the oscilloscope.

6. If multiple channels are used, then the current output from the different phases can be compared.

Although unable to compare individual phases, it is far safer to use a diagnostic scan tool to measure current to and from a motor generator. Live data can be graphed and flight-recorded while the vehicle is being driven under load and compared with the expected values. The test should be performed during acceleration and braking, following legal driving requirements.

Voltage Test – Oscilloscope

Testing the voltage from a motor generator should be conducted with the vehicle powered down and the high-voltage system isolated. This way, the motor generator can be carefully rotated by hand and the voltage and current output manually regulated by adjusting the speed at which it is turned. This may require the bypassing of a transmission park mechanism or the ability to turn the crankshaft of a hybrid vehicle; always follow the manufacturer's instructions and procedures.

Figure 5.34 Motor Generator Voltage connections using a Breakout Box

High-Voltage Diagnosis and Testing

1 • Once the high-voltage battery has been isolated and capacitors discharged, conduct a test to confirm the absence of voltage.

2 • Gain access to the motor generator phase terminals and connect the oscilloscope voltage probes to all three phases.

3 • Carefully rotate the motor generator to obtain a waveform.

4 • With three channels displayed, the voltage output from each phase can be compared.

Resolvers

The use of electric motors in a drive system requires an extremely accurate and effective method that shows the continuous relationship of speed and position between the magnetic field created in the stator and the corresponding speed and position of the rotor. This is because if they were to become 'out-of-sync', it would potentially lead to the stalling of the rotor or even operation in the wrong direction.

Traditional automotive speed and position sensors, including inductive, MRE and Hall effect, normally only produce a signal when relative movement takes place between two components.
As the motor generators of an electric vehicle start and stop during operation, these types of sensors are unsuitable to continuously monitor the position.

Resolvers, however, due to their design, are able to monitor speed and position regardless of whether the motor generator is rotating or stationary. This is because their design is a form of rotating transformer, which produces a constant output due to an input signal.
If the motor generator is moving, the output signal varies, but if stationary, a paused signal is still available.
The most common form of resolver used for motor generator speed and position detection takes advantage of variable **reluctance** (VR), which describes how hard it is for a magnetic force to go through something. Variable reluctance changes depending on the shape and type of the thing, and if there are any gaps or components that are not magnetic in it. The lower the reluctance, the easier it is for the magnetic force to go through. The higher the reluctance, the harder it is for the magnetic force to go through.

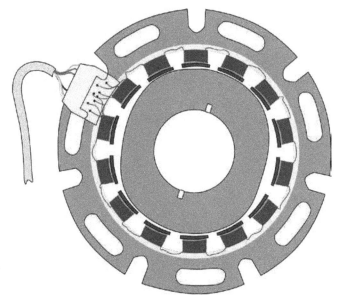

Figure 5.35 A Resolver

Its construction consists of two main components, a stator, and a rotor.
The stator is constructed from a number of coils, mounted on a frame which is aligned with the motor generator stator unit. The coils will be arranged so that one part will produce a continuous excitation signal/sinewave generated by an electronic control unit, and two sensor pick-up coils wound in opposite directions positioned at 90-degree intervals around the stator unit. The rotor of the resolver is fixed to the rotor of the motor/generator. It will be shaped so that as it turns within the resolver stator section, the air gap between the exciter and sensor coils will vary.

High-Voltage Diagnosis and Testing

The stable continuous **sinewave** produced by the **exciter** has its electromagnetic signature transferred to the pickup sensor coils via the shaped metal rotor through variable reluctance, resulting in two output signals, one with a sinewave signature and one with a cosine signature.

As the motor generator turns, the varying air gap created by the resolver rotor will change the amplitude of the output signal in both pickup coils, resulting in signals which essentially mirror each other as sine and cosine waves.

By using the exciter as the reference point, when comparing the amplitude of both the sinewave and cosine wave outputs, the potential difference between the two when compared to the excitation signal can be used to determine the exact position of the rotor in relation to the stator.

The speed signal is normally determined by assessing the time periods produced from the output pickups. As each time period relates to a particular amount of rotation (often one quarter turn), by dividing the periods into the amount of time elapsed, the rotational speed is obtained.

The number of time periods will relate to the shape of the resolver rotor.

Resolver designs using Hall effect generators and sensors are available and have been used for many years in different types of rotating machinery. However, due to the harsh environment of an electric vehicle transmission system and the sensitive nature of these types of sensors, a variable reluctance (VR) resolver using a type of transformer coil is often considered more robust.

Resolver Operation and Testing – Oscilloscope

Although this test is conducted on a low-voltage system at the resolver, precautions should be taken when working on or around high-voltage electrical systems, including the correct use of high-voltage personal protective equipment (PPE).

When testing a resolver, a **differential measurement** needs to be taken across each of the inductive coils. This means that oscilloscopes using a common ground are unsuitable for this form of test. Only oscilloscopes with a **floating ground** can be used, or those connected with a differential probe.

Depending on the configuration of the vehicle being tested, hybrid or fully electric, the engine may have to be operated while stationary or the vehicle may need to be appropriately raised and safely supported so that the driving wheel can be turned by hand. Always take appropriate precautions depending on which type of test is being conducted.

Access to manufacturer information will be required to help identify the resolver signal circuits and exciter. Connect three oscilloscope channels using a floating ground across the exciter coil, and both signal receiver coils.

Observing all precautions, the vehicle will need to be placed in ready mode and the motor generator rotated.

High-Voltage Diagnosis and Testing

Reluctance - the property of opposing the passage of magnetic flux lines.

Sinewave - a smooth, periodic oscillation that is mathematically described by the sine function.

Exciter - a device, often a small generator or a battery, that supplies the electric current used to produce the magnetic field in another generator or motor.

Differential measurement - the process of measuring the difference between two points, pressures, or electrical potentials.

Floating ground - a reference point for electrical potential in a circuit that is galvanically isolated from the actual earth ground.

When the motor generator is stationary, the following waveform should be obtained.

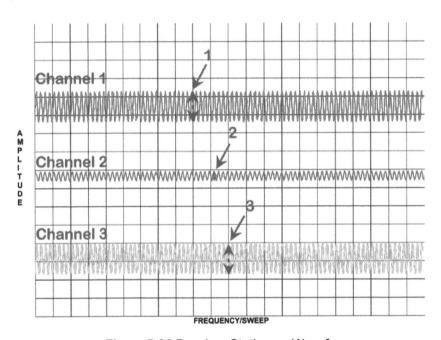

Figure 5.36 Resolver Stationary Waveform

Table 5.9 Waveform analysis resolver stationary	
Waveform component	**Description**
1	**Channel 1.** This section of the waveform shows a paused version of a sinewave from one of the output pickup coils. Depending on the position where the motor comes to a stop, its amplitude (height) will be frozen until the resolver starts to rotate again.
2	**Channel 2.** This section of the waveform shows a paused version of a cosine wave from one of the output pickup coils. Depending on the position where the motor comes to a stop, its amplitude (height) will be frozen until the resolver starts to rotate again.
3	**Channel 3.** This is the exciter waveform produced by the motor management system. This waveform remains constant while the vehicle is in ready mode, and does not change, regardless of whether the resolver is stationary or rotating.

High-Voltage Diagnosis and Testing

When the motor generator is turning, the following waveform should be obtained.

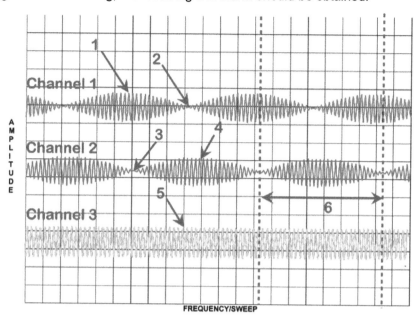

Figure 5.37 Resolver Turning Waveform

Table 5.10 Waveform analysis resolver rotating	
Waveform component	Description
1	**Channel 1.** This point on the waveform shows when the air gap between the rotor and the stator sinewave pickup coil is small, creating a proportionally sized amplitude (height) of the wave while the motor generator is revolving.
2	**Channel 1.** This point on the waveform shows when the air gap between the rotor and the stator sinewave pickup coil is large, creating a proportionally sized amplitude (height) of the wave while the motor generator is revolving.
3	**Channel 2.** This point on the waveform shows when the air gap between the rotor and the stator cosine wave pickup coil is small, creating a proportionally sized amplitude (height) of the wave while the motor generator is revolving.
4	**Channel 2.** This point on the waveform shows when the air gap between the rotor and the stator cosine wave pickup coil is large, creating a proportionally sized amplitude (height) of the wave while the motor generator is revolving.
5	**Channel 3.** This is the exciter waveform produced by the motor management system. This waveform remains constant while the vehicle is in ready mode, and does not change, regardless of whether the resolver is stationary or rotating.
6	**Channel 2.** This section of the waveform on channel 2 shows a period of rotation from the resolver. Depending on the resolver rotor shape, this will represent a portion of 360° rotation. (For example, an oval rotor will represent 90° of motor rotation.) The frequency of these periods will relate to the motor speed.

High-Voltage Diagnosis and Testing

 In reality, a combination of all three waveforms produced in the resolver is used to accurately calculate the speed and position of the motor generator.

 When diagnosing the waveforms from a motor generator resolver, if either or both of the signal coil pattern(s) are missing, then potentially the stator of the resolver has failed and will require replacement.

If the exciter signal and both of the signal coil patterns are missing, the issue could be caused by a faulty exciter coil, wiring, or exciter feed from the drive system control unit.

If the patterns are very noisy or messy, the issue may be caused by electromagnetic interference, damaged wiring, poor connections, or high resistance.

Incorrect alignment between the resolver's stator and rotor will provide incorrect position, leading to drive issues, or motor generator performance diagnostic trouble codes (DTC).

To help determine the location of the issue, tests should be conducted an both the ECU output and the reluctor input.

Inversion and Rectification

The process of converting direct current (DC) into alternating current (AC) is known as inversion.

The process of converting alternating current (AC) into direct current (DC) is known as rectification.

Batteries use direct current, while motor generators use alternating current. Therefore, it is important for a conversion process to take place between these two units. This is the job of the inverter. The inverter contains several components that perform inversion and rectification.

Capacitors

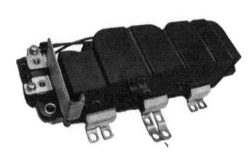

Housed inside the inverter, capacitors are used as a temporary storage for electrical energy. They act as a buffer or 'shock absorber' between the battery and the motor generators.
This reservoir of energy can be used to supply the high-power demands required by vehicle drive motors without placing excessive load on the high-voltage battery and act as a DC Bus filter.
They can also be used to smooth out irregularities or ripples in the generated electricity when recharging the high-voltage battery.

Figure 5.38 Capacitor Block

High-Voltage Diagnosis and Testing

IGBTs

Insulated gate bipolar transistors (IGBTs) are a form of very fast electronic switch with no moving components; [see Chapter 1]. They are connected in a circuit and switched in pairs in order to achieve inversion, which is the process of turning direct current into alternating current. They can be combined with diodes in a rectifier circuit in order to achieve conversion from alternating current to direct current when recharging.

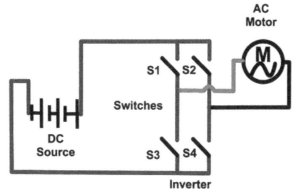

Figure 5.39 A Basic Inverter Circuit

An electronic circuit is created so that four switches (in this case the IGBTs) can change the route of electric current through a load, alternating the flow from one direction to another as shown in **Figures 5.40**. For simplicity, the IGBTs are shown as open or closed switches in the following illustrations.

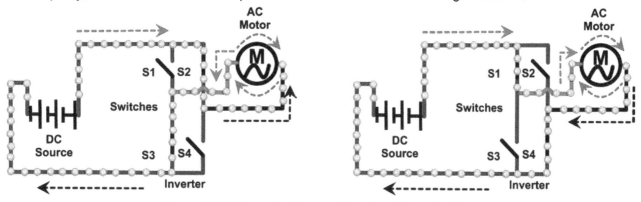

Figure 5.40 Changing Current Direction Using Switches

Although this is effectively alternating the current flow through the circuit, simplistic switching such as this will create a very crude square wave and not the smooth sinewave that is needed to efficiently operate the vehicle motor drive system.

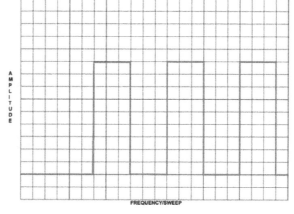

Figure 5.41 A Square Wave

High-Voltage Diagnosis and Testing

If a controller is used to modulate the switching pulse time by rapidly opening and closing the switches multiple times per cycle, an average rising and falling voltage can be obtained. This closely resembles a smooth sine wave.

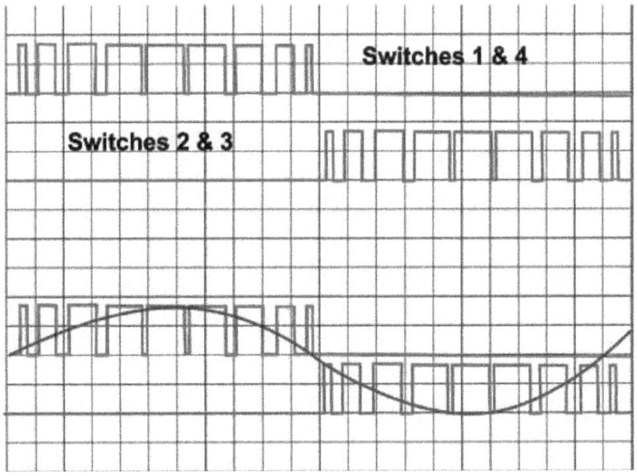

Figure 5.42 Modulated Switching

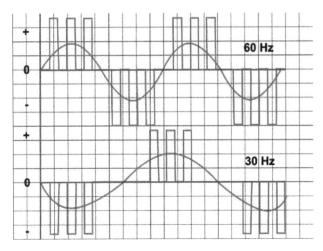

The inverter circuit is also able to vary the output voltage by controlling how long the switches are closed and current is flowing. This is known as pulse width modulation (PWM).

Figure 5.43 Voltage Control With PWM

The frequency of the waveform can be varied by controlling the timing of the switches. This is known as frequency modulation (FM).

Figure 5.44 Frequency Modulation

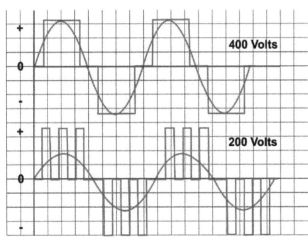

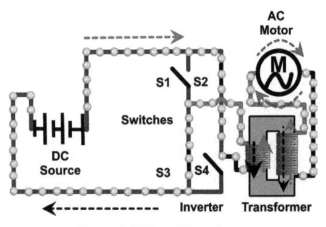

To increase the output to a level suitable for use with the vehicle drive motors, the AC is fed to a boost transformer. A boost transformer steps up the voltage via primary and secondary coils.

Figure 5.45 Boost Transformer

High-Voltage Diagnosis and Testing

The electric drive motors are three-phase, meaning that three circuits need to be connected to the switching mechanism, providing AC electric feed at 120-degree intervals. This creates a rotating magnetic field that drives the motor shaft.

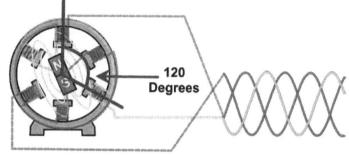

Figure 5.46 Three Phase

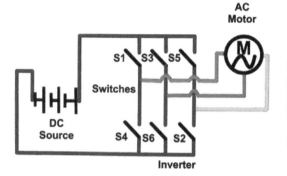

In order to provide three-phase AC electricity, six IGBT switches are connected in the pattern shown in the following illustrations and operated in sequence. This is known as a three-phase inverter bridge.

Figure 5.47 Three Phase Inverter Bridge

The cycle of switching through the phases is shown in the following illustrations:

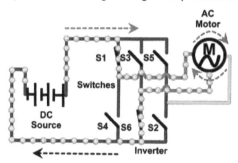

Figure 5.48 Switches 1 and 6 are Closed, Providing Phase 1 to Phase 2

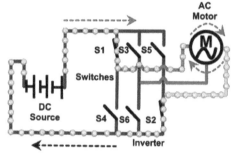

Figure 5.49 Switches 1 and 2 are Closed, Providing Phase 1 to Phase 3

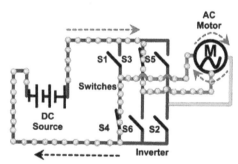

Figure 5.50 Switches 3 and 4 are Closed, Providing Phase 2 to Phase 1

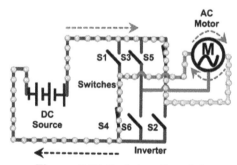

Figure 5.51 Switches 5 and 4 are Closed, Providing Phase 3 to Phase 1

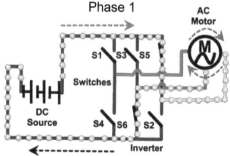

Figure 5.52 Switches 5 and 6 are Closed, Providing Phase 3 to Phase 2

This cycle repeats over and over, providing the three-phase electricity required for the drive motor windings. This is known as pulse train modulation (PTM).

High-Voltage Diagnosis and Testing

Inverter Thermal Management

IGBTs handle large amounts of electric current and voltage, and if not thermally managed, would quickly burn out and fail. Most vehicle inverters will have a coolant-fed heat exchanger mounted against the IGBT and other high-voltage/current components in order to regulate temperature. If the vehicle is a hybrid, this cooling system is likely to be independent of the one used for engine systems. An electrically driven pump is able to circulate coolant through the heat exchanger, which in turn may then circulate through an external radiator to dissipate excess heat to the atmosphere.

Inverter diagnosis is best conducted using the vehicle's onboard diagnostic system, first through stored codes and then via live data where possible. Physical measurements of the inverter unit itself while in situ are difficult due to its high-voltage nature.
Instead, a process of in and out should be used to help confirm symptoms.
If the DC feed and input signals are present and correct, then the output AC should also be correct. If not, suspect a failed inverter unit.

Another indication of inverter failure is smell. If components inside an inverter have burned out, by removing access covers, it is sometimes possible to smell the heat damage created. This is especially significant if the thermal management has failed due to coolant leakage or blockage, for example.

If you suspect that an inverter unit has failed, it may be prudent to dismantle it and conduct a visual inspection for damage to prove your diagnosis. However, this is not recommended if you intend to reuse the inverter or any of its internal components. Dismantling the inverter may void its warranty or cause further damage to the sensitive electronics inside.

Although components inside an inverter are often replaceable, once a failure diagnosis is confirmed, a complete replacement unit is recommended for repairs. The high-voltage nature means that many workshops will not have the facility to exchange internal components or test for correct function and safety. This has the potential to affect warranties or invalidate insurance in some cases as aftermarket repairs cannot provide the same level of guarantees as the original equipment manufacturer (OEM).

DC-to-DC Converter Operation and Testing

The DC-to-DC converter is a step-down transformer that is used to provide the low-voltage power supply for the vehicle's auxiliary system and maintain charge in the low-voltage auxiliary battery.

Sometimes contained within the inverter casing, but more often in its own separate unit, its operation can often be checked by measuring the low-voltage output auxiliary feed. This should be conducted at the low-voltage battery, but also at the DC-to-DC converter's auxiliary connection to the main system fuses.

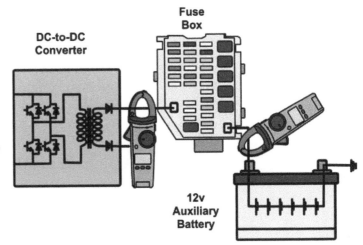

Figure 5.53 Low Voltage Measurements and Current Using a Clamp Meter

High-Voltage Diagnosis and Testing

When the vehicle is not in ready mode, the auxiliary system is maintained by the low-voltage battery. Once the vehicle is placed in ready mode, the DC-to-DC converter takes over the main system load, stepping high-voltage from the main drive battery down to a low auxiliary voltage. (For example, a 12 volt system may have an operating voltage of approximately 14 volts when the DC-to-DC converter is functioning.) As low-voltage wiring is often not shielded, an inductive current clamp can be connected around the wiring and assessed for current draw due to load or current feed due to charging.

The function and operation of a DC-to-DC converter can often be observed when testing the low-voltage system using an oscilloscope. For example, when testing the operation of the system main relays and contactors once the vehicle has engaged and progressed into ready mode, the function and operation of the DC-to-DC converter can be seen as a voltage increase after the contactors 12 volt switching. The interference created once the DC-to-DC converter has engaged is the result of the operation of the transformer using alternating current as part of the step-down process.

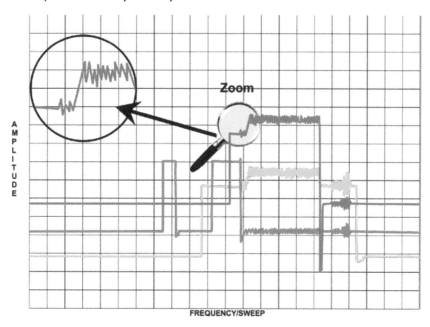

Figure 5.54 Zoomed View of the SMR Waveform Showing the Operation of the DC-to-DC Converter

HVAC Operation and Testing

PTC Heater Operation and Testing

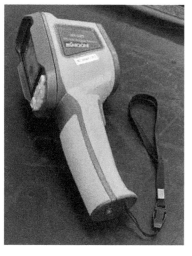

The cabin heating of an electric vehicle is often provided by a positive temperature coefficient (PTC) heater. This is an electric heating element that can be used to either warm air or coolant. The heat can then be transferred into the passenger compartment. An electrical failure in the operation of the PTC heater may often be indicated by a diagnostic trouble code. As this heating system is supplied from the high-voltage electrical circuit, any testing should be conducted with care, following any high-voltage precautions including personal protective equipment (PPE). The physical operation of the PTC heater can often be assessed by using an infrared thermometer or camera.

Figure 5.55 Infrared Camera

High-Voltage Diagnosis and Testing

AC Compressor Operation and Testing

The operation of the heating ventilation and air conditioning system (HVAC) of an electric vehicle will often require the use of a high-voltage electrically driven compressor; [see Chapter 3]. The diagnosis and testing of high-voltage compressors may be difficult due to the restricted nature of access to electrical connections, switching mechanisms and in some cases on-board inverters.

Live data should be used to confirm the operation of the air conditioning compressor and any further diagnosis or repairs should only be undertaken if the technician is suitably trained and qualified to meet environmental and legal requirements.

Similar to a PTC heater, the physical operation of the heating ventilation and air conditioning system can often be assessed by using an infrared thermometer or camera.

Electric Vehicle Supply Equipment (EVSE) Operation and Testing

Unless the electric vehicle uses a hydrogen fuel cell as its portable power source, electric vehicles will require a method of externally charging the high-voltage battery. This is achieved using an electric vehicle supply equipment (EVSE). [See Chapter 4 for charging mode methods and charging plug/socket types].

A limited amount of testing and diagnosis can be conducted on the charging operation between the EVSE and vehicle, however, this may require the use of a dedicated electric vehicle supply equipment breakout box.

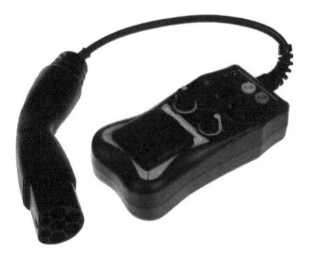

Figure 5.56 EVSE Breakout Box

It is important to remember that when a vehicle is connected to the electric vehicle supply equipment (EVSE) to charge, the high-voltage vehicle system will be live and awake. Therefore, all high-voltage precautions will need to be observed, including the use of correctly rated and calibrated electrical test equipment and high-voltage personal protective equipment (PPE).

In order to conduct any in-vehicle testing of mains charging systems using an oscilloscope, a differential probe with a minimum of a Category 3 (CAT III) rating will be required. A standard attenuator is not suitable and does not provide the correct protection for both the operator and the scope or vehicle.

High-Voltage Diagnosis and Testing

Proximity Pilot (PP) Operation and Testing

The proximity pilot (PP) is one of the connections used as part of the communication protocol between the vehicle's onboard charger (OBC) and the charging cable. It performs a number of roles depending on the type of charging cable connected to the electric vehicle. It uses a fixed resistor inside the charging cable plug, which, when plugged-in to the vehicle charging socket, indicates connection.

The proximity pilot circuit will communicate the connection status of the charging cable, the current carrying capability in amps of the charging cable, and initialise immobiliser activation, which stops the vehicle from entering ready mode and being driven if a charging cable is connected.

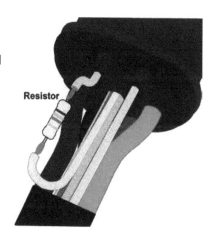

Figure 5.57 Proximity Pilot Resistor

The proximity pilot provides a discrete circuit between the vehicle's onboard charger and charging cable. It does not connect between the electric vehicle and the EVSE.

A small voltage is supplied to the PP circuit from the onboard charger, and the measured current flow can then be used to determine the size of the fixed resistor in the charger plug and therefore the connection status and cable capacity.

The proximity pilot circuit can be tested using an ohmmeter. Access to manufacturer specific data, wiring diagrams and information will be required in order to conduct this test.

1
- Dismantle and gain access to the vehicle's onboard charger unit (OBC).

2
- Locate the wiring harness containing the proximity pilot (PP) communication wire and disconnect from the OBC.

3
- Calibrate the ohmmeter by touching the two probes together, and then place one probe on the proximity pilot loom connection and one on the OBC proximity circuit ground.

4
- With the charging cable disconnected, the reading on the ohmmeter should show off limits (O/L) or infinity.

5
- Ensuring that the charge cable is not connected to any mains electricity, insert the charge plug into the vehicle's charging socket. Read and record the resistance value obtained and compare with the manufacturer's specification.

A resistance test using an ohmmeter requires that the components or circuit to be tested are powered down, disconnected and isolated from the rest of the vehicle. This may involve working on or in close proximity to the high-voltage systems. Where possible this should be shut down and isolated if necessary, following all specific high-voltage safety precautions including personal protective equipment (PPE).

High-Voltage Diagnosis and Testing

 **SAFETY** When testing the proximity pilot (PP) circuit of a charging cable, it must never be connected to mains electricity or an EVSE unit. Never test a tethered cable as this is permanently connected to a charger and therefore may attempt to begin the charging process if plugged into a vehicle.

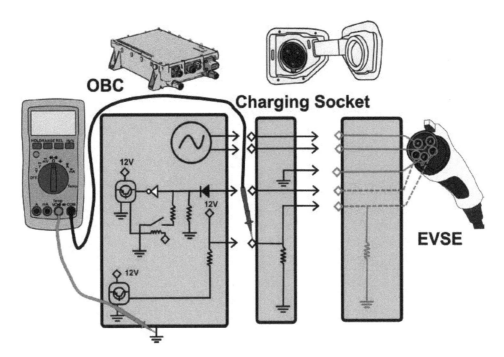

Figure 5.58 Testing the Proximity Pilot Resistance

Approximate values of resistance and cable current capacity are shown in Table 5.11.

Table 5.11 Proximity Pilot resistance values	
Resistance	**Charge cable current capacity**
Type 1 cable	
2700 Ohms Ω (2.7 kΩ)	Cable unplugged
145 Ohms Ω	Cable plugged in
411 Ohms Ω	Cable release trigger pressed
Type 2 cable	
Off limits (O/L or infinity)	Cable unplugged
1500 Ohms Ω (1.5 kΩ)	13 Amps
680 Ohms Ω	20 Amps
200 Ohms Ω	32 Amps
100 Ohms Ω	70 Amps (single phase) 63 Amps (three phase)

High-Voltage Diagnosis and Testing

Control Pilot (CP) Operation and Testing

The control pilot (CP) is used for bidirectional communication between the EVSE and the electric vehicle's onboard charger (OBC). The communication consists of an amplitude modulated (voltage) and pulse width modulated (PWM) signal to convey charging system readiness/state and charging current.
- The amplitude (voltage) indicates the charging stage.
- The pulse width modulation (duty cycle) indicates the charge delivery.

It is possible to test the communication between the control pilot (CP) and the electric vehicle supply equipment (EVSE) using an oscilloscope with a differential probe. Special precautions must be taken, as the high-voltage onboard charger will be exposed while the vehicle is connected to the mains outlet. This means that live circuits at high-voltage have the potential to cause injury, death, vehicle or test equipment damage. Do not attempt any charging system tests unless you have received adequate training and are using appropriate high-voltage personal protective equipment (PPE).

In order to conduct the following test, the vehicle's high-voltage battery should be partially discharged so that when the EVSE charge cable is connected to the vehicle, communication and charging will commence.

Ensuring that all health and safety procedures are followed, including the use of appropriate personal protective equipment (PPE), dismantle and gain access to the vehicle's on-board charger unit.
Use the manufacturer's test data to identify the on-board charger's control pilot connection.
Attach a differential probe with a minimum CAT III rating to one of the oscilloscope channels and select an appropriate scale on the differential probe switch (see the manufacturer's recommendations).
Attach the connections of the differential probe to the circuit as follows:
- Signal probe to the control pilot (CP) connection at the on-board charger (OBC).
- Negative probe to the vehicle chassis or negative low-voltage auxiliary battery terminal.

With the oscilloscope running, connect the charging cable from the EVSE to the vehicle, capture and analyse a waveform.

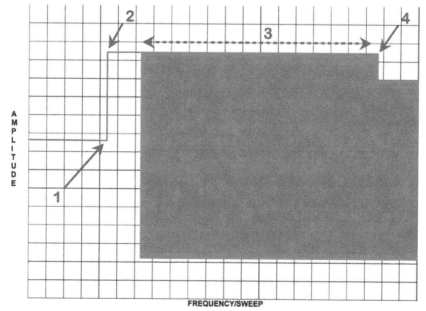

Figure 5.59 CP Connected Communication Waveform

High-Voltage Diagnosis and Testing

Table 5.12 CP connected communication	
Waveform component	Description
1	This is the point on the waveform where the EVSE charge cable is disconnected and the voltage on the control pilot (CP) line is at 0 volts.
2	This is the point on the waveform where the charging cable is connected and the voltage on the control pilot (CP) line jumps to approximately 9 volts.
3	This section shows the start of bidirectional communication between the vehicle and the EVSE, lasting for about 6 seconds. This communication uses a pulse-width modulation with a frequency of around 1,000 Hz and an amplitude modulation of between +9 volts and -12 volts. [The communication duty cycle is too fast to view without zooming in, see **Table 5.13**].
4	This point on the waveform shows where charging has started, and the voltage drops to a modulated amplitude between +6 volts and -12 volts. Bidirectional communication continues and the duty cycle of the pulse width varies with the current delivered. [The communication duty cycle is too fast to view without zooming in, see Table **5.14**].

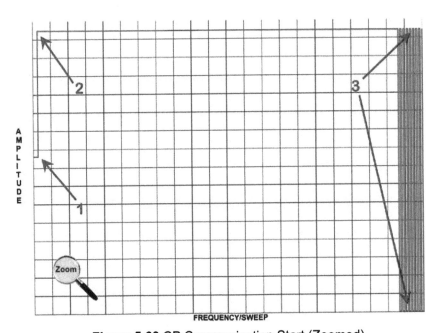

Figure 5.60 CP Communication Start (Zoomed)

Table 5.13 CP connected communication Start (zoomed)	
Waveform component	Description
1 (Zoom)	This is the point on the waveform where the EVSE charge cable is disconnected and the voltage on the control pilot (CP) line is at 0 volts. (Zoomed in).
2 (Zoom)	This is the point on the waveform where the charging cable is connected and the voltage on the control pilot (CP) line jumps to about 9 volts. (Zoomed in).

High-Voltage Diagnosis and Testing

Table 5.13 CP connected communication Start (zoomed)	
3	This section shows the start of bidirectional communication between the vehicle and the EVSE, lasting for about 6 seconds. This communication uses a pulse-width modulation with a frequency of around 1,000 Hz and an amplitude modulation of between +9 volts and -12 volts. (Zoomed in).

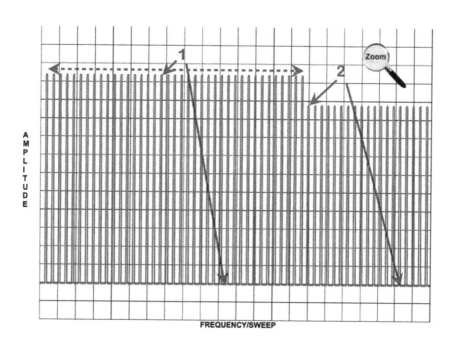

Figure 5.61 CP Charging Start (Zoomed)

Table 5.14 CP connected charging start (zoomed)	
Waveform component	**Description**
1	This section shows the start of bidirectional communication between the vehicle and the EVSE, lasting for about 6 seconds. This communication uses a pulse-width modulation with a frequency of around 1,000 Hz and an amplitude modulation of between +9 volts and -12 volts. (Zoomed in).
2	This point on the waveform shows where charging has started, and the voltage drops to a modulated amplitude between +6 volts and -12 volts. Bidirectional communication continues and the duty cycle of the pulse width varies with the current delivered. (Zoomed in). [See **Table 5.16**]

High-Voltage Diagnosis and Testing

Table 5.15 indicates the charging states from the EVSE depending on approximate voltage amplitude.

\multicolumn{5}{|c|}{Table 5.15 EVSE charging state}				
Charging cable connected Yes/No	Charging status	Approximate voltage	Charging Possible	Description
No	Standby	0 volts	No	EVSE not connected to the vehicle.
Yes	Vehicle connected	9 volts	No	EVSE connected to the vehicle but not yet charging.
Yes	Charging	6 volts	Yes	EVSE connected to the vehicle, charging in progress.
Yes	Ventilation	3 volts	Yes	Ventilation in progress, charging fans running.
Yes	EVSE Shutdown	0 volts	No	EVSE fault or proximity pilot (PP) short to Earth.
Yes	Error	-12 volts	No	EVSE unavailable.

Duty cycle of the PWM signal and approximate current delivery are shown in Table 5.16.

\multicolumn{2}{|c|}{Table 5.16 Relationship between duty cycle and charge current}	
Duty cycle percentage (%)	Approximate charge current
10%	6 Amps
20%	12 Amps
30%	18 Amps
40%	24 Amps
50%	30 Amps
66%	40 Amps
80%	48 Amps
90%	65Amps
94%	75 Amps
96%	80 Amps

High-Voltage Diagnosis and Testing

Summary

This chapter has described:

- Safety precautions and procedures when working on EV/Hybrid vehicles.
- High and low voltage batteries including analysis.
- High-voltage battery diagnosis and testing.
- Contactors and SMR operation and testing.
- Vehicle operating information and CAN Bus.
- Motor generator operation and testing.
- Inversion and rectification.
- DC to DC converter operation and testing.
- HVAC operation and testing.
- Electric vehicle supply equipment (EVSE) operation and testing.

Common Acronyms/Abbreviations

This section contains common acronyms and abbreviations: An acronym is a word that is formed from the first letters of a phrase or a series of words, usually to make it easier to say or remember. This list is not exhaustive, but provides some acronyms used in conjunction with the design and operation of electric vehicles. Abbreviations may have different meanings or designations depending on context, and acronyms may be further adapted, reused, or reinterpreted as technology and engineering develops.

A - Amperes
A/F - Air Fuel Ratio
A/T - Automatic Transmission
AAT - Ambient Air Temperature
ABS - Antilock Brake System
AC - Alternating Current
AC - Air Conditioning
ACC - Automatic Climate Control
ACC - Air Conditioning Clutch
ACR - Air Conditioning Relay
ACR4 - Air Conditioning Refrigerant, Recovery, Recycling, Recharging
ADU - Analogue-Digital Unit
AED - Automatic Electronic Defibrillators
AEV- All Electric Vehicle
AFR - Air Fuel Ratio
AGM- Absorbed Glass Matt
Ah- Amp Hours
AM - Amplitude Modulation
ATS - Air Temperature Sensor
AVD - Aqueous Vermiculite Dispersant
AVO- Amps Volts Ohms
AWG - American Wire Gage
BBW - Brake by Wire
BCM - Body Control Module
BCM - Battery Control Module
BEV - Battery Electric Vehicle
BHP - Brake Horsepower
BMS - Battery Management System
BMU - Battery Management Unit
BOB - Breakout Box
BPP - Brake Pedal Position Switch
BTS - Battery Temperature Sensor
Btu - British thermal unit
BUS N - Bus Negative
BUS P - Bus Positive
C - Celsius
CA - Cranking Amps
CAN - Controller Area Network
CAT - Category
CC - Catalytic Converter
CC - Climate Control
CC - Cruise Control
CC - Cubic Centimetres
CCA - Cold Cranking Amps
CCS - Combined Charging System
CCV - Closed Cricut Voltage

CFC - Chlorofluorocarbons
CL - Closed Loop
CLV - Calculated Load Value
CNG - Compressed Natural Gas
CO - Carbon Monoxide
CO2 - Carbon Dioxide
COSHH - Control of Substances Hazardous to Health
CP - Control Pilot
CPU - Central Processing Unit
CRC - Cyclic Redundancy Check
CTP - Closed Throttle Position
CTS - Coolant Temperature Sensor
CV - Constant Velocity
CVT - Continuously Variable Transmission
DBW - Drive by Wire
DC - Duty Cycle
DC - Direct Current
DMM - Digital Multimeter
DLC - Data Link Connector (OBD)
DSO - Digital Storage Oscilloscope
DTC - Diagnostic Trouble Code
EBCM - Electronic Brake Control Module
EBD - Electronic Brake Force Distribution
ECC - Electronic Climate Control
ECM - Engine/Electronic Control Module
ECS - Emission Control System
ECT - Engine Coolant Temperature
ECU - Electronic Control Unit
EECS - Evaporative Emission Control System
EEGR - Electronic EGR (Solenoid)
EEPROM - Electronically Erasable Programmable Read Only Memory
EGO - Exhaust Gas Oxygen Sensor
EGR - Exhaust Gas Recirculation
EGRT - Exhaust Gas Recirculation Temperature
EMF - Electromotive Force (voltage)
EMI - Electromagnetic Interference
EML - Engine Management Light
EOBD - European Onboard Diagnostics
EPA - Environmental Protection Act
EPB - Electronic Parking Brake
EPB - Equipotential bonding
EPROM - Erasable Programmable Read Only Memory
EPS - Electronic Power Assisted Steering
ESP - Electronic Stability Programme
ESS - Engine Start-Stop
EV - Electric Vehicle
EVAP - Evaporative Emissions System
EVAP CP - Evaporative Canister Purge
EVSE - Electric Vehicle Supply Equipment
EWS - Immobiliser (network acronym)
FM - Frequency Modulation
FOT - Fixed Orifice Tube
FSD - Full Scale Deflection
FWD - Front Wheel Drive
GND - Electrical Ground Connection
GWP - Global Warming Potential

Common Acronyms/Abbreviations

H - Hydrogen
HASAWA- Health and Safety at Work Act
H2O - Water
HC - Hydrocarbons
HCA - Hot Cranking Amps
HEGO - Heated Exhaust Gas Oxygen Sensor
HEV - Hybrid Electric Vehicle
HFC - Hydrogen Fuel Cell
HFC - Hydrofluorocarbon
HFO - Hydrofluoroolefin
HICE - Hydrogen Internal Combustion Engine
HO2S - Heated Oxygen Sensor
hp - Horsepower
HSE - Health and Safety Executive
HT - High Tension
HV - High Voltage
HVAC - Heating Ventilation and Air Conditioning
HVIL - High Voltage Interlock Loop
Hz - Hertz
I/O - Input / Output
IA - Intake Air
IAT - Intake Air Temperature
IC - Integrated Circuit
ICCB - In Cable Charging Box
ICE - In Car Entertainment
ICE - Internal Combustion Engine
IGBT - Insulated Gate Bipolar Transistor
IGN - Ignition
IHKA – Climate Control (network acronym)
ISO - International Standard of Organisation
KAM - Keep Alive Memory
Kg/cm2 - Kilograms/Cubic Centimetres
KHz - Kilohertz
Km - Kilometres
Kombi - Instrument Cluster (network acronym)
KPA - Kilopascal
KWP - Keyword Protocol
l - Litres
LA - Lead Acid
LCD - Liquid Crystal Display
LED - Light Emitting Diode
LEV - Low Emission Vehicle
LFP - Lithium Iron Phosphate
LHD - Left Hand Drive
Li-ion - Lithium ion
LMO - Lithium Manganese Oxide
LOS - Limited Operating Strategy
LPG - Liquefied Petroleum Gas
LTO - Lithium Titanate
LWB - Long Wheelbase
LWR - Vertical Headlight Control (network acronym)
M/T - Manual Transmission
MAC - Mobile Air Conditioning
MCM - Motor Control Module
MEF - Methane Equivalency Factor
MF - Maintenance Free
MIL - Malfunction Indicator Lamp
MPG - Miles per Gallon
MPH - Miles per Hour
MRE - Magnetic Resistive Element
MRS - Multiple Restraint System (network acronym)
mS or ms - Millisecond
MSD - Maintenance Service Disconnect/Manual Service Disconnect
mV or mv - Millivolt
N - Nitrogen
NCA - Nickel Cobalt Aluminium
NCAPS - Non-Contact Angular Position Sensor
NCM - Nickel Cobalt Manganese
NCRPS - Non-Contact Rotary Position Sensor
NGV - Natural Gas Vehicles
NIB - Neodymium Iron Boron
Ni-MH - Nickel Metal Hydride
Nm - Newton Meters
NOx - Oxides of Nitrogen
NPN - Negative Positive Negative
NTC - Negative Temperature Coefficient
O2 - Oxygen
OBC - Onboard Charger/Offboard Charger
OBD I - On Board Diagnostics Version I
OBD II - On Board Diagnostics Version II
OCV - Open Circuit Voltage
OD - Outside Diameter
ODP - Ozone Depletion Potential
OE - Original Equipment
OEM - Original Equipment Manufacturer
OFN - Oxygen Free Nitrogen
OL - Off Limits
OL - Open Loop
OS - Oxygen Sensor
P/N - Part Number
PAG - Polyalkylene Glycol
PATS - Passive Anti-Theft System
PCB - Printed Circuit Board
PCM - Powertrain Control Module
Pd - Potential Difference (volts)
PE - Protected Earth
PEF- Propane Equivalency Factor
PEM - Proton Exchange Membrane
PEV - Pure Electric Vehicles
PH - Potential Hydrogen
PHEV - Plug-in Hybrid Electric Vehicle
PID - Parameter Identification Location
PKE - Passive Keyless Entry
PLC - Powerline Communication
PNP - Positive Negative Positive
POE - Polyolester Oil
POF - Plastic Optical Fibre
POT - Potentiometer
PP - Proximity Pilot
PPE - Personal Protective Equipment
PPM - Parts Per Million
PPS - Accelerator Pedal Position Sensor
PROM - Programmable Read-Only Memory
PSI - Pounds per Square Inch
PTC - Positive Temperature Coefficient

Common Acronyms/Abbreviations

PTM - Pulse Train Module

PUWER - Provision and Use of Work Equipment Regulations
PWM - Pulse Width Modulation
RAM - Random Access Memory
RBS - Regenerative Braking system
RCD - Residual Current Device
RCM - Reserve Capacity Minutes
RCM - Restraint Control Module
RDS - Radio Data System
RDW -Tyre Pressure Monitoring (network acronym)
RE EV - Range Extended Electric Vehicles
REF - Reference
RESS - Rechargeable Energy Storage System
RFI - Radio Frequency Interference
RHD - Right Hand Drive
RIDDOR - Reporting of Injuries Diseases and Dangerous Occurrence Regulations
RKE - Remote Keyless Entry
RMS - Recovery Management Station
ROM - Read Only Memory
RON - Research Octane Number
RTV - Room Temperature Vulcanizing
RWD - Rear Wheel Drive
SAE - Society of Automotive Engineers (Viscosity Grade)
SEI - Solid Electrolyte Interphase/Interface
SIPS - Side Impact Protections System
SMR - System Main Relay
SoC - State of Charge
SoH - State of Health
SRI - Service Reminder Indicator
SRS - Supplementary Restraint System (air bag)
SRT - System Readiness Test
SWB - Short Wheelbase
SWL- Safe Working Load
SZM - Central Switch Module (network acronym)
TACH - Tachometer
TCM - Transmission Control Module
TCS - Traction Control System
TP - Throttle Position
TPM - Tyre Pressure Monitor
TPP - Throttle Position Potentiometer
TPS - Throttle Position Sensor
TSB - Technical Service Bulletin
TXV - Thermal Expansion Valve
UART - Universal Asynchronous Receiver-Transmitter
UJ - Universal Joint
ULEV - Ultra Low Emission Vehicle
USB - Universal Serial Bus
UV - Ultraviolet
V - Volts
V2G - Vehicle to Grid
VAC - Vacuum
VDU - Visual Display Unit
VDE - Verband der Elektrotechnik

VIN - Vehicle Identification Number
VPE - Vehicle Protection Equipment

VR - Variable Reluctance
VSS - Vehicle Speed Sensor
W/B - Wheelbase
Wh - Watt Hours
WPT - Wireless Power Transfer
WSS - Wheel Speed Sensor
WVO - Waste Vegetable Oil
ZEV - Zero Emission Vehicle

Appendix

The information contained in this section, especially legislation, regulations and standards, is subject to continuous review and reform. Please ensure that you are always working in accordance with the latest information and guidelines.

Health and safety legislation

Legislation is the process of making or enacting laws. Specific and non-specific legislation will always need to be observed and followed; however, the legislation will vary depending on your geographical location. It is important that you are aware of the legislation and your rights and responsibilities, as well as those of your employer. It is your right to expect your employer to fulfil their responsibilities and it is your employer's right to expect you to fulfil yours. Legislation is the law and, if you do not observe it, you are committing an offence.

Some examples of UK work-based legislation are listed below:

1 The Health and Safety at Work etc. Act 1974: defines the general duties of everyone from employers and employees to owners, managers, and maintainers of work premises for maintaining health and safety within most workplaces.

2 The Electricity at Work Regulations 1989: require people in control of electrical systems to ensure they are safe to use and maintained in a safe condition.

3 The Personal Protective Equipment at Work Regulations 1992: places a duty on every employer in Great Britain to ensure that suitable PPE is provided to employees who may be exposed to a risk to their health or safety while at work.

4 The*Provision and Use of Work Equipment Regulations 1998: place the responsibility for the safety of workplace equipment on anyone who has control over the use of work equipment, including your employer, you and your colleagues.

5 The Control of Substances Hazardous to Health Regulations 2002: requires employers to assess the risks that arise from the use of hazardous substances. This will include any arrangements to deal with accidents, incidents, or emergencies, such as those resulting from serious spillages.

6 The Management of Health and Safety at Work Regulations 1999: requires employers to carry out risk assessments, put in place measures to minimise risks, appoint competent people and arrange for appropriate information and training for their staff.

7 The Workplace (Health, Safety and Welfare) Regulations 1992: cover a wide range of basic health, safety, and welfare issues such as ventilation, heating, lighting, workstations, seating and welfare facilities.

8 The Health and Safety (Display Screen Equipment) Regulations 1992: set out requirements for work with computers and visual display units (VDUs).

9 The Manual Handling Operations Regulations 1992: cover the moving of objects by hand or bodily force.

10 The Health and Safety (First Aid) Regulations 1981: cover the requirements for first aid, including the number of trained first aiders required in the workplace.

11 The Health and Safety Information for Employees Regulations 1989: requires employers to display posters telling employees what they need to know about health and safety.

12 The Employers' Liability (Compulsory Insurance) Act 1969: requires employers to take out insurance against work-related accidents and ill health involving employees and visitors to the premises.

13 The Noise at Work Regulations 1989: requires employers to take action to protect employees from hearing damage.

Appendix

14 The Pressure Safety Systems Regulations 2000: require users and owners of pressure systems to demonstrate that they know the safe operating limits, principally pressure and temperature, of their pressure systems, and that the systems are safe under those conditions. This will include compressed air systems but does not include pressurised systems used for vehicle propulsion.

15 Reporting of Injuries, Diseases and Dangerous Occurrences Regulations 1995: requires employers to report accidents, near misses and ill health to the HSE and to keep records of these events.

Standards and regulations

The production, maintenance and operation of electric vehicles must conform to standards and regulations in order to provide a level of quality and safety. These standards and regulations cover not only the vehicle itself, but also the tools and equipment associated with maintenance and repairs.

Some examples of standards and regulations are shown below:

ECE R100 is a regulation that specifies the safety requirements for the electric powertrain of road vehicles, including rechargeable battery systems. It is issued by the United Nations Economic Commission for Europe (UNECE) and applies to vehicles which are mainly passenger cars and commercial vehicles. The regulation covers various aspects of the electric power train, such as electrical safety, fire resistance, mechanical integrity, thermal shock, vibration, overcharge, over-discharge, and over-temperature protection.

UN 38.3 is a regulation that specifies the safety requirements for the transport of lithium batteries, which are classified as dangerous goods because of their potential fire hazard. The regulation applies to lithium metal and lithium-ion batteries, whether they are shipped alone or installed in a device. The regulation requires the batteries to pass a series of tests, such as altitude simulation, temperature cycling, vibration, shock, external short circuit, impact, overcharge, and forced discharge, before they can be transported by air, sea, rail, or road.

IEC 62660 is a series of international standards that specify the performance, reliability and safety requirements of secondary lithium-ion cells for the propulsion of electric vehicles. The series consists of three parts:
- IEC 62660-1: Performance testing.
- IEC 62660-2: Reliability and abuse testing.
- IEC 62660-3: Safety requirements.

ISO 6469-1 is an international standard that specifies the safety requirements for the rechargeable energy storage systems (RESS) of electrically propelled road vehicles, such as battery-electric vehicles, fuel-cell vehicles, and hybrid electric vehicles. The standard covers the protection of persons inside and outside the vehicle and the vehicle environment from potential hazards caused by the RESS, such as electrical shock, fire, mechanical damage, thermal shock, vibration, overcharge, over-discharge and over-temperature. The standard also specifies the test methods and criteria for evaluating the safety performance of the RESS.

ISO 12405-4 is an international standard that specifies the test specification for lithium-ion traction battery packs and systems for electrically propelled road vehicles, such as hybrid electric vehicles, plug-in hybrid electric vehicles, battery electric vehicles and fuel cell vehicles. The standard covers the performance, reliability and electrical functionality of the battery packs and systems, such as capacity, power, energy, efficiency, no load SoC loss, cranking power, internal resistance, and cycle life. The standard also specifies the test methods and criteria for evaluating these characteristics.

ISO 19453-6 specifies requirements for lithium-ion traction battery packs or systems used in battery electric, hybrid electric and fuel cell electric road vehicles. It describes the most relevant environmental stresses and specifies tests and test boundary conditions. It also establishes a classification of battery packs or systems and defines different stress levels for testing when a classification is applicable and required.

Appendix

American standards

FMVSS 305 stands for Federal Motor Vehicle Safety Standard No. 305, which is a regulation issued by the National Highway Traffic Safety Administration (NHTSA) in the United States. The standard specifies performance requirements for limitation of electrolyte spillage and retention of electric energy storage/conversion devices during and after a crash, and protection from harmful electric shock during and after a crash and during normal vehicle operation.

SAE J2464 describes a set of tests that can be used to evaluate the response of RESS to conditions or events that are beyond their normal operating range, such as overcharging, overheating, short circuiting, crushing, penetrating, or burning. The tests are intended to cover a broad range of vehicle applications and RESS technologies, and to provide a common framework for comparing the performance of different RESS.

SAE J2929 is a safety standard for electric and hybrid vehicle propulsion battery systems that use lithium-based rechargeable cells. It defines the minimum requirements for the design, construction, testing, and certification of the battery systems and their components, such as cells, modules, packs, and subsystems. The standard aims to ensure the safety, reliability, and performance of the battery systems in various conditions, such as normal operation, crash, and abuse.

Chinese standards

GB/T 31484 is a Chinese national standard that specifies the cycle life requirements and test methods for traction batteries of electric vehicles. The standard covers different types of traction battery technologies, such as lithium-ion, nickel-metal hydride, lead-acid, and supercapacitors. The standard aims to ensure the safety, reliability, and performance of the traction battery in various conditions, such as normal operation, crash, and abuse.

GB/T 31486 is a Chinese national standard that specifies the electrical performance requirements, test methods, and inspection rules for traction battery of electric vehicle. The standard covers different types of traction battery technologies, such as lithium-ion, nickel-metal hydride, lead-acid, and supercapacitors. The standard aims to ensure the safety, reliability, and performance of the traction battery in various conditions, such as normal operation, crash, and abuse.

GB/T 38032 is a Chinese national standard that specifies the safety requirements and test methods for electric buses. It applies to electric and hybrid electric buses which are vehicles with more than eight seats for passengers, excluding the driver's seat. The standard covers various aspects of electric bus safety, such as waterproof and dustproof performance, fireproof performance, rechargeable energy storage system safety, control system safety, charging safety, and safety after vehicle crash and rollover.

GB/T 18384 is a Chinese national standard that specifies the safety requirements and test methods for electric vehicles. It applies to electric vehicles whose maximum working voltage of the on-board drive system is class B voltage, which is more than 60 VDC or 30 VAC. The standard covers various aspects of electric vehicle safety, such as protection of persons against electric shock, functional safety protection, traction battery safety, collision protection, flame-retardant protection, charging interface safety, vehicle alarm and prompt, vehicle event data recording, and electromagnetic compatibility.

Tools and equipment

IEC 60900 is a standard that applies to hand tools that are used for working live or close to live parts at nominal voltages up to 1000 VAC and 1500 VDC. The standard aims to ensure the safety, reliability, and performance of the tools and to protect the users from electric shock and short-circuiting.

IEC/EN 60903 is a standard that applies to electrical insulating gloves and mitts that provide protection of the worker against electric shock. The standard also covers electrical insulating gloves with additional integrated mechanical protection, referred to as "composite gloves". The standard aims to ensure the safety, reliability, and performance of the tools and to protect the users from electric shock and short-circuiting.

Appendix

IEC 61111 is an international standard that applies to electrical insulating matting made of elastomer for use as a floor covering for the electrical protection of workers on electrical installations. The standard aims to ensure the safety, reliability, and performance of the matting and to protect the users from electric shock and short-circuiting.

Charging standards

IEC 62196 is a series of international standards that define the requirements and tests for plugs, socket-outlets, vehicle connectors and vehicle inlets for conductive charging of electric vehicles. The standards aim to ensure the safety, reliability, and performance of the charging equipment and to protect the users from electric shock and short-circuiting.

IEC 63110 is an international standard that defines a protocol for the management of electric vehicles charging and discharging infrastructures.

DIN SPEC 70121 is a German technical specification that defines the digital communication between a direct current (DC) electric vehicle charging station and an electric vehicle for control of DC charging in the Combined Charging System (CCS).

GB/T 18487.1 is a Chinese national standard that specifies the general requirements for electric vehicle conductive charging systems. It covers the classification, size, shape, weight, technical requirements, test methods, inspection rules, packaging, and marking of the charging systems and their components, such as plugs, sockets, connectors, inlets, and cables. It also covers the safety issues, such as electric shock, overload, short circuit, and emergency stop, as well as the communication and connection issues between the electric vehicle and the charging station. The standard applies to both alternating current (AC) and direct current (DC) charging systems with different voltage levels and power ratings.

GB/T 20234 is a series of Chinese national standards that define the requirements and tests for connectors and interfaces for conductive charging of electric vehicles. The standards cover both alternating current (AC) and direct current (DC) charging systems with different voltage levels and power ratings. The standards aim to ensure the safety, reliability, and performance of the charging equipment and to protect the users from electric shock and short-circuiting.

GB/T 27930 is a Chinese standard for electric vehicle battery charging. It defines the communication protocols between the off-board conductive charger and the battery management system (BMS) in the vehicle. The standard is based on the SAE J1939 network protocol and uses the CAN bus with a point-to-point connection. The standard covers various aspects of the charging process, such as electrical safety, fire resistance, mechanical integrity, thermal shock, vibration, overcharge, over-discharge, and over-temperature protection.

SAE J1772 is a standard for electrical connectors for electric vehicles in North America. It specifies the physical, electrical, communication and safety requirements for the electric vehicle conductive charge system and coupler. It allows electric vehicles to be charged from various sources of electric power, such as household outlets, public charging stations, or renewable energy sources.

ISO 15118 is an international standard that defines a vehicle to grid (V2G) communication interface for bi-directional charging and discharging of electric vehicles. The standard provides various use cases such as secure communication, smart charging and the Plug & Charge feature that allows an electric vehicle to automatically identify and authorise itself to a compatible charging station. The standard covers both conductive and wireless power transfer technologies and applies to different types of electric vehicles, such as battery-electric vehicles, fuel-cell vehicles, and hybrid electric vehicles.

SAE J2954 is a standard for wireless power transfer (WPT) for electric vehicles led by SAE International. It defines three classes of charging speed, WPT 1, 2 and 3, at a maximum of 3.7 kW, 7.7 kW and 11 kW, respectively.

Appendix

SAE J2847/2 is a standard that defines the communication between plug-in electric vehicles and off-board DC chargers. It specifies the requirements and specifications for the electric vehicle conductive charge system and coupler, which uses the SAE J1772 connector. It also enables the communication between the electric vehicle and the charging station to ensure safe and efficient charging.

IEC 61980 is an international standard that specifies the safety, performance, and communication requirements for electric vehicle wireless power transfer (WPT) systems. WPT systems use electromagnetic fields to transfer electric power from a supply device to an electric vehicle without physical contact. The standard applies to WPT systems that operate at standard supply voltages up to 1000 VAC and up to 1500 VDC and cover various types of electric vehicles, such as battery-electric vehicles, fuel-cell vehicles and hybrid electric vehicles.

Index

2

2-Pole Tester · 29

A

Absorbed Glass Mat · 119
AC and DC - Three Phase · 11
Accidents · 72, 76
Acid · 93
Active · 178, 180, 188
Additional Multimeter Functions · 27
Air Conditioning · 67, 100, 101, 143, 224, 225, 237
Alessandro Volta · 6, 9
Alkaline · 93
alternating current · 12, 18, 19, 22, 65, 66, 68, 70, 96, 105, 127, 137, 138, 139, 140, 141, 142, 143, 147, 152, 154, 173, 188, 203, 210, 211, 215, 230
Alternating Current · 12, 24
Alternative Propulsion · 44, 69
Ambient temperature · 173
Ammeter · 26
Ammonia · 47, 48, 71
Ammonia Green · 47
Amp Clamp · 31
Amperage · 52
Ampere-hour · 118
Amps Clamp · 30
Anaerobic · 45
Analogue · 23, 62, 224
André-Marie Ampère · 9
Anode · 163
Anthropogenic · 40
Aqueous · 83, 126, 224
Aqueous Vermiculite Dispersion · 83
Arc blast · 95, 104, 108, 156
Arc flash · 86, 95, 104, 108, 110, 156, 158
Armature · 136
Atom · 6, 7, 159
Atoms and Molecules · 6
Attenuators · 34
Automatic Defibrillator · 115
Autoranging · 23
Auxiliary · 62, 119
Axial Flux · 137, 140

B

Balance · 177
Balancing · 167, 171, 177, 178, 181, 185
Batteries · 16, 63, 117, 159, 164, 210
Battery · 8, 13, 16, 62, 65, 107, 117, 118, 119, 121, 122, 123, 124, 126, 128, 133, 143, 154, 160, 161, 163, 165, 166, 170, 171, 173, 177, 178, 182, 184, 185, 202, 224
Bio-alcohol · 47, 71
Biodegradable · 46
Biodiesel · 46
Biogas · 45
Biogenic · 45

BMS · 143, 170, 171, 172, 173, 177, 224, 230
BMU · 65, 167, 168, 224
Braided · 135
Brake-By-Wire · 97
Brush · 137
Brushes · 136
Burns · 91, 104, 108, 156
Bus bar · 173
Butanol · 43

C

Cables · 65, 95, 134
Cabling · 133, 134
Calorific value · 42
Capacitor · 17, 18, 121, 142, 188, 189, 190, 191, 210
Capacitors · 17, 77, 96, 188, 210
Capacity · 118, 160, 161, 226
Carbon dioxide · 39, 40, 43, 84, 94
Carbon footprint · 40
Carbon monoxide · 43, 84, 94
Carbon neutral · 46
Catalyst · 50, 145
Cathode · 163, 165
Caustic · 93
CCS · 68, 151, 153, 224, 230
Cell · 123, 166, 170, 183
Cetane · 47
CHAdeMO · 68, 151, 153
Charge Shunting · 179
Charge Shuttles · 179
Charging · 38, 59, 68, 71, 106, 123, 133, 147, 148, 150, 151, 152, 153, 161, 162, 181, 221, 222, 224, 225, 230
Chemical Reactions · 8
Chemical Risks · 126
Chemical Spills · 92
Chemistry · 93, 161, 165
Circuit · 8, 15, 16, 17, 129, 179, 189, 194, 195, 196, 211, 225
Circuit Protection · 17
Climate change · 40
Climate Change · 39
Closed-circuit voltage · 11
Commutator · 136
Compounds · 93
Compressed Natural Gas · 45, 224
Compressor · 67, 101, 143, 216
Condenser · 101
Conductor · 8
Connector · 134, 224
Contactors · 132, 155, 186, 187, 192, 193, 194, 223
Continuity · 8, 27, 134
Control of Substances Hazardous to Health Regulations · 81, 227
Control Pilot · 149, 219, 224
Controlled waste · 82, 93
Cooling · 99, 100, 124, 125
Copyright · 1, 2, 59
Corrosive · 93
Cranking Amps · 118, 224, 225
C-rate · 123, 165
Cross-sectional area · 95

Index

Current · 6, 9, 11, 12, 13, 14, 15, 18, 24, 26, 121, 136, 137, 141, 159, 168, 172, 180, 204, 205, 211, 214, 224, 226
Cylindrical · 166

D

Data link connectors · 35
DC-to-DC converter · 62, 65, 79, 80, 118, 128, 133, 142, 214, 215
DC-to-DC Converter · 65, 142, 214, 215
Destination Charger · 148
Diesel · 41, 42, 43, 47, 69
Differential measurement · 208
Differential Probe · 35
Digital Multimeter · 23, 24, 224
Diode · 21, 28, 142, 225
direct current · 12, 13, 18, 22, 29, 54, 65, 66, 68, 70, 80, 96, 117, 127, 136, 137, 139, 140, 141, 142, 143, 147, 152, 154, 173, 180, 188, 210, 211, 230
Direct injection · 50
Drivers Display · 56, 62
Drivetrains · 58
Dual · 59, 69

E

Earth return · 119
Earth Return · 119
ECE R100 · 104, 105, 228
ECE R-100 · 53
Ecotoxic · 82
Electric and Electronic Components · 18
Electric Circuits · 15
Electric shock · 91, 104, 108, 156
Electric Vehicle Supply Equipment · 66, 68, 216, 224
Electric Vehicles · 54, 109, 157, 225, 226
Electrical Tooling and Measurement Devices · 23
Electrical Units and Terminology · 9
Electrically Propelled Vehicles · 52
Electricity · 5, 8, 17, 25, 48, 227
Electrocution · 5, 38, 53, 72, 107, 155
Electrode · 18, 174
Electrolysis · 50, 144, 145
Electrolyte · 93, 115, 118, 121, 126, 174, 226
Electromagnet · 137
Electromagnetic interference · 135
Electromagnetism · 8
Electromotive force · 10, 19
Electromotive Force · 9, 14, 224
Electron · 7, 8
Electrons · 7, 146
EMF · 9, 10, 14, 15, 19, 117, 118, 135, 159, 180, 198, 224
Engine Bays · 57
Environmental Protection · 82, 111, 224
EOBD · 175, 183, 224
Equipotential bonding · 135, 224
Equipotential Bonding · 32
Ethanol · 43, 47, 71
Evaporator · 101
Exciter · 208
Exhaust Emissions · 42

Exothermic · 93, 126
Expansion Valve · 101, 103, 226
Explosion · 104, 108, 126, 156
Extended Range Electric Vehicles · 54

F

Faraday · 114, 163
Field coil · 136
Field effect · 137
Fire · 5, 42, 72, 82, 84, 104, 108, 113, 126, 156
Fire Blankets · 84
Fire Extinguishers · 82, 113
Fire Triangle · 42
First Aid · 85, 91, 114, 227
First Aid Box · 91
First Responder Loops · 78
Fixed Orifice Tube · 102, 103, 224
Flash point · 46
Floating ground · 208
Flyback · 180
Fossil fuel · 42
Frequency · 28, 29, 137, 212, 224, 226
Fuel Cell · 54, 145, 146, 225
Full Hybrids · 57
Fuses · 17

G

Gear Selector · 56
Generator · 16, 66, 96, 203, 205
Generators · 8, 16, 66, 96, 136
Georg Ohm · 10
Global warming · 39, 40
Global Warming Potential · 40, 224
Greenhouse effect · 40
Greenhouse gas · 50

H

Hall Effect · 22, 172, 173
Hand Tools · 116, 129
Hazard · 72, 74, 76, 80, 104, 106, 111
Hazard Management · 74, 76
Hertz · 29, 225
horsepower · 10, 70
HVIL · 64, 75, 112, 225
Hybrid · 1, 38, 53, 55, 56, 58, 60, 61, 63, 64, 65, 69, 72, 96, 104, 107, 108, 136, 143, 154, 155, 156, 223, 225
Hybrid and Electric Vehicles · 1, 38, 53
Hydrocarbon · 41
Hydrocarbons · 43, 84, 94, 225
Hydroelectric · 48
Hydrogen · 6, 42, 43, 49, 50, 51, 52, 54, 70, 84, 94, 144, 145, 146, 225

I

IGBTs · 140, 211, 214
Immobilisation · 77

Index

Induction · 19, 137, 139
Inductive · 30, 31, 173, 180
Inductors · 19
Inert · 84
Information · 72, 73, 106, 107, 108, 129, 155, 156, 183, 197, 227
Infotainment · 62
Infrastructure · 153
Insulated Floor Mat · 116
Insulated Return · 133, 135
insulated-gate bipolar transistor · 22
Insulation · 32, 116, 131, 172, 173, 183
Insulator · 8
Internal Combustion Engines · 50
Inversion · 155, 210, 223
Inverter · 65, 96, 139, 140, 211, 213, 214
Inverters · 65
Ions · 118
Isolated · 130
Isolation · 32, 78, 128, 172, 173

J

Jacking Points · 94
James Watt · 10
Jump Starting · 79

K

Kilowatt · 10

L

Lead Acid · 63, 117, 161, 170, 225
Legislation and Regulations · 80
Light-emitting diode · 21
Li-ion · 122, 123, 162, 164, 225
Liquified Petroleum Gas · 44
Lithium · 63, 122, 123, 126, 162, 165, 170, 174, 225
Lithium-Ion · 63
Litmus paper · 93
Live Data · 132, 175
Lockout · 114, 182
Lorentz force · 22
Lossless · 181
Luigi Galvani · 6

M

Magnetic fields · 53
Magnetic Flux · 19
Magnets · 8
Maintenance · 1, 64, 72, 75, 92, 97, 98, 225
Mandatory · 73, 74
Megohmmeter · 31, 32, 131
Membrane · 146, 225
Mennekes · 68, 150
Methane · 40, 43, 45, 225
Methanol · 43, 47
Micro Hybrids · 57
Mild Hybrids · 57

Milliamps · 95
Milliohm Meter · 33, 203
Milliohm meters · 33, 204
Modes · 61, 68, 147
Molecule · 6
Molecules · 145
MOSFET · 22
Motor · 13, 18, 66, 96, 136, 155, 203, 205, 223, 225, 229
Motors · 8, 18, 66, 136, 137, 138, 139, 140
Movement of Electrons · 7
MSD · 64, 75, 225
Multimeter · 23, 24, 25, 26, 27, 116
Multimeters · 23
Multi-point injection · 50
Mutual induction · 153

N

Network · 197, 198, 200, 224
Neutralised · 93
Nickel Cadmium · 121
Nickel Metal Hydride · 63, 121, 170, 171, 225
Nickel-Metal Hydride · 121
Nominal voltage · 123, 163
North American Standard · 152
NTC · 172, 173, 225
Nucleus · 6

O

Octane · 47, 226
Ohmmeter · 25
Ohms and Power Law · 14
Ohms Law · 14
On-Board Chargers · 66
Open circuit · 11, 119
Open circuit voltage · 11
Organic matter · 45
Oscillating · 153
Oscilloscope · 33, 73, 108, 156, 194, 200, 204, 205, 207, 224
Oscilloscopes · 33, 237
Oxides of nitrogen · 44
Oxygen · 6, 224, 225

P

Parallel · 15, 16, 58, 59, 69, 124, 167, 168, 169
Parasitic drain · 80, 121
Particle cracking · 174
Passive · 177, 188, 225
Peak oil · 40, 42
Periodic Table of Elements · 7
Personal Protective Equipment · 64, 72, 85, 106, 225, 227
Petrol · 41, 43, 69
pH · 92, 93
Phase · 139, 203
Photodiode · 21
Plug-in Hybrid Electric Vehicles · 53
Plug-in Hybrids · 57

Index

Polarity · 80, 121, 136
Pollution · 42
Polymath · 163
Polymer gel · 163
Positive Temperature Coefficient · 67, 225
Potential Difference · 9, 10, 201, 225
Potentiometers · 20
Pouch · 114, 166
Power · 10, 13, 14, 15, 16, 52, 58, 69, 161, 174, 175, 190, 191, 224, 226
Power Law · 15
power-split · 58
Pre-charge · 189, 190, 191, 194, 195, 196
Pre-ignition · 50
Prismatic · 166
Prohibition · 73, 74
Propagate · 93
Propane · 43, 225
Proton · 8, 225
Protons · 146
Proving Unit · 29, 127
Proximity Pilot · 149, 217, 218, 225
PTC Heater · 67, 144, 215
PTC Heating · 67
Pulse width modulation · 137
Pure Electric · 54, 225
Pyrolysis · 50
Pyrotechnic · 18

Q

Quadricycles · 54

R

Ready Mode · 60, 131
Reasonably practicable · 80
Receiver Drier · 101
Recovery Position · 91, 92
Rectification · 141, 142, 210
Redox Shuttle · 171, 173, 181
Refinery · 41
Refrigeration Cycle · 102
Regenerative braking · 59, 61
Regenerative Braking · 60, 97, 226
Relay · 18, 224, 226
Relays · 18, 132, 187
Reluctance · 19, 208, 226
Reserve Capacity Minutes · 118, 226
Residual Current Devices · 18
Resistance · 10, 13, 14, 25, 52, 163, 186, 192, 203, 218
Resistor · 20, 217
Resistors · 20
Resolvers · 206
Responsibilities · 105
Responsibilities of Management and Skilled Employees · 105
Rheostats · 20
Risk · 80, 104, 105
Risk Assessments · 80
Roadside Recovery · 78

S

Safety Rescue Hook · 114
Safety Rescue Pole · 90
Safety Signs · 73
Safety Systems · 73, 228
Scame · 150
Scan Tools · 35
Schottky diode · 21
Semiconductor · 8
Series · 15, 16, 58, 59, 69, 124, 167, 168, 169
Series and Parallel Circuits · 15
Shielding · 134, 135
Short circuit · 5, 53, 94, 126
Shunt · 173
Signs and Barriers · 113
Sinewave · 208
Single-point injection · 50
Smart key · 62
SMR · 75, 79, 132, 155, 186, 188, 189, 190, 191, 192, 193, 194, 195, 196, 215, 223, 226
Solar · 48, 70
Solid electrolyte interphase · 174
Solid State · 163
Spill Kit · 115, 126
SRS · 75, 78, 226
State of charge · 122, 170, 183
State of Charge · 128, 170, 226
State of health · 170, 173, 183
static electricity · 6, 89
Stator · 137
Suction Accumulator · 102
Supercharger · 148
Suppressant · 84
Suppressants · 82, 84
Synchronous · 137, 138
Syngas · 48, 50
Synthetic · 48, 50

T

Temperature · 29, 67, 73, 108, 156, 172, 173, 183, 224, 225, 226
Terminal · 134
The Discovery of Electricity · 6
Thermal breakdown · 94
Thermal explosion · 127
Thermal Management · 124, 125, 214
Thermal runaway · 84, 93, 126, 127
Thermistors · 20, 122, 124
Three Phase · 11, 12, 139, 213
Three-Phase · 67, 96
Torque · 130
Towing · 78
Toxic · 93
Transformer · 19, 180, 212
Transformers · 19
Transistor · 22, 29, 142, 225
Transistors · 21, 22, 29, 141
Trickle charge · 122
Tunnel diode · 21
Tyres · 98

Index

U

Ultra-Low Emission Vehicle · 52

V

Vehicle to Grid · 154, 226
Vermiculite · 83, 84, 224
Volatile · 126
Volt Drop · 193
Voltage · 1, 9, 11, 13, 14, 15, 24, 26, 31, 52, 65, 75, 77, 78, 95, 107, 113, 116, 117, 119, 123, 128, 130, 131, 132, 133, 134, 143, 155, 159, 165, 170, 171, 182, 194, 204, 205, 212, 214, 224, 225

W

Washing · 76
Waste Transfer Note · 82
Waterproofing and Submersion · 75
Watts · 10, 15, 55
Wave power · 48
What is Electricity · 6
Why Do We Need Electric Vehicles · 39
Wind power · 48
Wireless · 153, 226

Y

Yazaki · 68, 149

Z

Zener diode · 21
Zero Emission Vehicle · 52, 226

If you enjoyed this book, please check out other titles by the author:

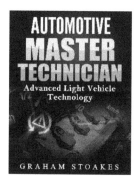

Paperback: 338 pages
Publisher: Graham Stoakes
(1 Feb. 2015)
ISBN-10: 099294922X
ISBN-13: 978-0992949228

Level 4 Automotive Master Technician - Advanced Light Vehicle Technology

'Technology needs technicians, and the ability to harness technical diagnosis calls for a Master Technician'.

The rapid growth in technology used in the production of cars has highlighted the need for a different approach to vehicle diagnosis and repair. The integration of complex electronic control with mechanical systems shows the brilliance in the engineering capabilities of designers and manufacturers.

While this technology has improved the comfort, safety, convenience and reliability of vehicles, it has also created an issue with established methods of maintenance and repair. As many of the control systems operate beyond our natural capabilities, diagnostic tooling is required to undertake most of the fault finding duties traditionally conducted by vehicle technicians. Also, the sophisticated nature of advanced system faults will often lead to diagnostic requirements for which there is no prescribed method.

One of the fundamental roles of a Master Technician will be the diagnosis and repair of these complex and advanced system faults, for which diagnostic approaches need to be developed that can provide logical strategies to reduce overall diagnostic time. An effective diagnostic routine should always begin with a logical assessment of symptoms and then uses reasoning to reduce the possible number of options, before following a systematic approach to finding and fixing the root cause.

Paperback: 154 pages
Publisher: Graham Stoakes
(6 July 2015)
ISBN-10: 0992949246
ISBN-13: 978-0992949242

Principles of Light Vehicle Air Conditioning

'As the number of vehicles on the world's roads rises, the demand for increased levels of comfort and convenience also grows'

While air conditioning and climate control may be seen as a luxury by some, the key benefits often outweigh the initial costs and resources required to implement these systems on newly produced vehicles; in fact most new cars come with some form of air conditioning as standard.

An environment which helps keep the driver and passengers comfortable and alert, maintaining the correct levels of ventilation and humidity, can increase concentration and the ability to devote more of their attention to the occupation of driving.

The downside of these systems is the environmental impact of the chemicals used to provide the refrigeration process.

Globally, anthropogenic, or 'man-made' emissions are believed to be the key factor in climate change and refrigerants have a larger influence than many others.

Small amounts of fluorinated gasses released to atmosphere may be causing irreparable damage to our planet, initiating ozone depletion and global warming.

Although many organisations are currently seeking alternatives to these harmful cocktails, at the present time we are restricted by the availability, cost and technology required to make viable replacements.

This means that for the time being, technicians and air conditioning professionals need to ensure that refrigerants are handled with due diligence and systems are maintained to the highest standards in order to contain and reduce emissions. Remember these chemicals only become dangerous when released to atmosphere.

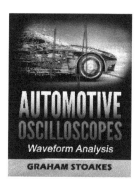

Paperback: 198 pages
Publisher: Graham Stoakes
(24 April 2017)
ISBN-10: 099294262
ISBN-13: 978-0992949266

Automotive Oscilloscopes Waveform Analysis

The rapid growth of technology used in cars has highlighted the need for a piece of diagnostic equipment that will give you X-ray vision and show you the heartbeat of a vehicles electrical and electronic system.

An OBD scan tool is vital for modern vehicle diagnostics; however, trouble codes will only take you so far. The problem can arise when the phrase 'fault code' is used in connection with diagnosis. A code will rarely point you directly to the root cause of a vehicle fault but can help you focus your diagnosis on a specific area and run functional tests.

It is the oscilloscope (or scope) that can truly test the operation and health of a system component. The important thing to remember about a oscilloscopes, is that they should be easy to set up and use; otherwise they will be passed over for a more familiar tool within your comfort zone.

Remember that nothing ever happens within your comfort zone.

There is a great deal of misconception about how difficult a scope can be to set up, and once you are used to your own equipment, if it is laid out and ready to use, it will soon become your diagnostic tool of choice.

This book has been written to help you get the most from your oscilloscope and has been designed to give straightforward and uncomplicated methods that can be used effectively for automotive diagnosis. It covers many of the most common automotive waveforms, assisting you in the analysis of the patterns produced, without restricting you to rigid equipment settings, or vehicle system design.

This will give you the 'scope' to develop your systematic diagnostic routines, with the flexibility to adapt to changing requirements.

www.grahamstoakes.com

Printed in the USA
CPSIA information can be obtained
at www.ICGtesting.com
CBHW040423030924
14008CB00031B/1607